Communication Research into the Digital Society

Fundamental Insights from the Amsterdam School of Communication Research

Edited by
Theo Araujo and
Peter Neijens

Amsterdam University Press

Cover design: Coördesign, Leiden
Lay-out: Crius Group, Hulshout

ISBN 978 90 4856 059 2
e-ISBN 978 90 4856 060 8
DOI 10.5117/9789048560592
NUR 610

Printed and bound by CPI Group (UK) Ltd, Croydon, CR0 4YY

Communication Research into the Digital Society

Table of Contents

Preface

Media and communication have become ubiquitous in today's societies and affect all aspects of life. On an individual level, they impact how we learn about the world, how we entertain ourselves, and how we interact with others. On an organisational level, the interactions between media and organisations, such as political parties, NGOs, businesses and brands, shape organisations' reputation, legitimacy, trust and (financial) performance, as well as individuals' consumer, political, social and health behaviours. At the societal level, media and communication are crucial for shaping public opinion on current issues such as climate change, sustainability, diversity, and well-being. Media challenges are widespread and include mis- and disinformation, the negative impact of algorithms on our information diets, challenges to our privacy, cyberbullying, media addiction, and unwanted persuasion, among many others. All this makes the study of media and communication crucial.

This book provides a broad overview of the ways in which people create, use, and experience their media environment, and the role of media and communication for individuals, organisations, and society. The chapters in the book were written by researchers from the Amsterdam School of Communication Research (ASCoR) on the occasion of its 25th anniversary. ASCoR is today the largest research institute of its kind in Europe and has developed over the past 25 years into one of the best communications research institutes in the world.

The book is organised into two parts. In the first part, we reflect on the development of the research institute against the background of developments in academia over the past 25 years. We also discuss how the dynamic communication environment and critical media challenges have inspired ASCoR's research over the past quarter of a century.

In the second part, the book provides an overview of empirical findings as well as theoretical and practical contributions from studies conducted at ASCoR. Each of the four research groups at ASCoR is represented: Political Communication & Journalism, Youth & Media Entertainment, Corporate Communication, and Persuasive Communication. The book also includes cross-cutting topics such as human–machine communication, artificial intelligence, person-specific media effects, computational communication science, and a sociological approach towards communication, linking communication to meaning and innovation. We hope and trust that our academic colleagues, students, science journalists, policymakers, the general public, and other ASCoR stakeholders enjoy this rich overview of digital society communication research.

The book has benefited greatly from the suggestions and comments of our reviewers: Peter van Aelst (University of Antwerp), Wouter van de Brug (University of Amsterdam), Sandra Diehl (University of Klagenfurt), Karin Fikkers (University of Utrecht), Ine Goovaerts (University of Antwerp), Tilo Hartman (VU University Amsterdam), Claartje ter Hoeven (Erasmus University Rotterdam), Ciska Hoving (Maastricht University), Yan Jin (University of Georgia, USA), Christine Liebrecht (Tilburg University), Felicia Loecherbach (New York University, USA), Jörg Matthes (University of Vienna), and Brahim Zarouali (KU Leuven). Pieke Assmann, Yufang Arya Liao, Yuzhi Lu, Miriam van der Putte, Margriet Smit, Ania Tekień, Amber van der Wal, and Elizabeth Wang assembled data, conducted analyses, performed writing checks, and provided the much-needed administrative support to make this book possible. We thank them for their excellent contributions. We also thank the members of ASCoR's international advisory board for their invaluable contributions to our research institute. The board includes, since 2015, Amy Jordan (Rutgers University, USA), Frank Esser (University of Zurich, Switzerland), Robin Nabi (University of California, Santa Barbara, USA), Mette Morsing (Copenhagen Business School, Denmark), Dhavan Shah (University of Wisconsin-Madison), and Michael Slater (The Ohio State University, USA).

Part I.

ASCoR's History

1. How International, National, and Local Research Strategies Shaped ASCoR's History in Its First 25 Years

Peter Neijens, Sandra Zwier, Claes H. de Vreese, Jochen Peter, Rens Vliegenthart, and Theo Araujo[1]

Abstract
This chapter gives an overview of the institutional development of the Amsterdam School of Communication Research (ASCoR) in the 25 years since its foundation in 1997, paying attention to both objectives of ASCoR: organising and supporting research, and training of PhD students. The chapter also discusses various factors that shaped this trajectory, including ASCoR's orientation on media technological and societal changes, its international, funding, output, and research methods' orientation, and individual factors. The chapter concludes with ASCoR's new policy being deployed to address some concerns about the current research climate at universities in the Netherlands.

Keywords: institutional development, communication research history, research institute, research policy

Introduction

In only 25 years, the Amsterdam School of Communication Research (ASCoR), the research institute in communication science at the University of Amsterdam (UvA), has firmly established itself as one of the top

1 Peter Neijens, Claes H. de Vreese, Jochen Peter, Rens Vliegenthart, and Theo Araujo served successively as ASCoR's scientific directors. Sandra Zwier was ASCoR's first research manager. ASCoR's founding director, Jan van Cuilenburg, passed away in 2016.

Neijens, P., Zwier, S., de Vreese, C. H., Peter, J., Vliegenthart, R., & Araujo, T. (2024). How international, national, and local research strategies shaped ASCoR 's history in its first 25 Years. In T. Araujo & P. Neijens (Eds.), *Communication research into the digital society: Fundamental insights from the Amsterdam School of Communication Research* (pp. 11-31). Amsterdam University Press.
https://doi.org/10.5117/9789048560592_ch01

communication research institutes worldwide. This is evidenced by consistently high ratings by international review committees and the number 1 position in the QS World Ranking by Subject for the sixth consecutive year in 2023. ASCoR has become home to a large, impactful, and highly visible community of researchers, being "cutting edge in the topics it addresses, the methods it uses, and the theories it develops" (Research Review, 2020, p. 11).

In this chapter we provide a brief overview of the institute's history and show its developments over the years, paying attention to both objectives of ASCoR: organising and supporting research, and training of PhD students. We have taken great advantage of ASCoR's annual and self-assessment reports since 1997, prepared by ASCoR's directors and research managers. We used the data from these reports for our overview and adopted some verbatim descriptions of the research programme and mission.

The start of communication science at the University of Amsterdam

Communication science was established in the Faculty of Political and Social Sciences, which was founded after the Second World War. Kurt Baschwitz was the first professor of the discipline with an appointment for the field of "Press, Public Opinion and Propaganda." The great interest in these subjects was motivated by the developments in Nazi Germany in the preceding period (Hemels, 2017; van Ginneken, 2018; Wieten, 2005). After Baschwitz's promotion to full professor in 1952, he held the chair in "Press Science and Mass Psychology," which reflected a combination of the then dominant medium (the press) and mass psychology.

After an intensive effort for academic recognition of the field, the programmes for the degree in communication science were started, first at Radboud University in Nijmegen (in 1982) and then at the University of Amsterdam (in 1985) (Hemels, 2017). Until the launch of the study programme, subjects such as press studies, broadcasting studies, film studies, sociology of mass communication, public relations, and mass psychology were taught as electives in social sciences degree programmes (Hemels, 2017).

The new Communication Science study programme at the UvA was based on the two pillars that had already been part of Baschwitz's chair: Mass Communication (including press and television, chaired by Denis McQuail), and Mass Psychology (chaired by Marten Brouwer). Shortly after the start, a third pillar was added to the programme: the appointment of Jan van Cuilenburg in 1988 broadened the Communication Science programme with courses on media policy, media economics, information science, advertising, and public communication. Appendix 1.1 shows the chairs in communication science until the launch of ASCoR.

This history and position of communication science at UvA explains ASCoR's social scientific approach to the discipline. This approach is different from media and communication programmes that tend to emphasise language and cultural aspects and can often be found within faculties of humanities.

The foundation of ASCoR

The foundation of ASCoR was inspired by an increase in publications in international journals, research presentations at international conferences, active involvement and success in acquiring science foundation grants, and the founding of a PhD programme in communication research in the second half of the 1990s. The immediate reason for the founding of ASCoR were the conclusions of an external review committee (1990–1994), which rated the quality of communication research as "good to excellent" and unequivocally concluded that a research school in communication science should be implemented as soon as possible. Department Chair Jan van Cuilenburg then took the initiative to set up the new communication science research institute.

Van Cuilenburg became the director of the school "in formation." Together with two of the school's board members, Holli Semetko and Peter Neijens, and its research manager, Sandra Zwier, he spent 1997 developing the mission and profile of the institute that would be named The Amsterdam School of Communications Research *ASCoR*. Its mission statement is shown in Table 1.1.[2] ASCoR was officially launched with an opening conference on September 18 and 19, 1997,[3] at the Royal Netherlands Academy of Arts and Sciences (KNAW), with contributions from the minister of education of the Netherlands and international colleagues, including Robin Mansell (University of Sussex, UK), Peter Dahlgren (University of Lund, Sweden), John Keane (University of Westminster, UK), Thomas Petterson (Harvard University, USA), Rohan Samarajiva (Ohio State University, USA), Jean-Claude Burgelman (Free University Brussels, Belgium), Tamar Liebes (Hebrew University, Israel), and Jane Fulk (USC Annenberg School for Communication,

2 In 2009 "Communications" was replaced with "Communication."

3 See Appendix 1.2 for an overview of ASCoR's management and staff from 1997.

USA) (van Cuilenburg & Zwier, 1998). Members of parliament, government officials, publishers, and representatives of communication agencies and market research companies also attended the conference. On the second day, a valedictory ceremony in honour of Denis McQuail was held. His valedictory lecture "Theory as Compass: Competing Lodestones for Communication Research" was the concluding contribution to the conference.

Table 1.1. ASCoR's foundational mission statement in 1997

The Amsterdam School of Communications Research (ASCoR) conducts research at an advanced academic level into the political, social, psychological, cultural, and economic aspects of communications infrastructure, contents, and effects. Research addresses the role of media and (tele)communications, nationally and internationally, from the perspective of democracy and processes of opinion and identity formation. As an institute for academic research and training in the field of communications, ASCoR is also a meeting place for scholars, students, and policymakers. ASCoR aims at high-quality research into the fundamental aspects of communications, its social potential and effects. ASCoR's research is rooted in the empirical tradition and characterised by methodological pluralism. Research is aimed at theory development and testing based on data. ASCoR also addresses applied research questions in the field of communications as part of the school's societal task next to taking the opportunity to learn from real-life situations. Research in ASCoR also has an eye for normative and ethical issues. ASCoR takes the perspective of "communication toward open societies." This perspective refers to the situation where senders and receivers of communication may freely exchange messages. Openness in opinion formation is indicated by the processing of diverse information, opinions, and views which may contribute to tolerance of different cultures, views, and groups of people in society. ASCoR aims to be a truly international research organisation with respect to research topics, publication outlets, PhD students, exchange programmes, and personnel. At the same time, it is a strong wish of ASCoR to collaborate with all Dutch universities in the area of communication research.

ASCoR's mission and lines of research

Table 1.1 shows that it has been ASCoR's aim from the beginning to contribute knowledge on the production, uses, and consequences of information and communication in informing, persuading, and entertaining citizens, at the individual, group, institutional, and societal levels. Much of this remains in place today. Likewise, ASCoR is still guided by the conviction that research should provide answers to both fundamental scientific, and socially relevant questions, with its focus on an interdisciplinary, empirical approach.

The main lines of research by ASCoR scholars have since 2008 crystallised in the four themes shown in Table 1.2: corporate communication, persuasive communication, political communication and journalism, and youth and media entertainment. Popular culture was a line of research which, although very successful, unfortunately could not be continued due to the departure of the programme director, Liesbet van Zoonen, who left ASCoR for Loughborough University, UK. Appendix 1.3 lists ASCoR's research programmes and programme directors from 1997.

Table 1.2. ASCoR's lines of research

Corporate Communication focuses on the development, execution, and especially the effects of communication strategies towards internal and external stakeholders of organisations. In this context, this group focuses on mediatisation processes, artificial intelligence and technology use, legitimacy and visibility, and sustainability and diversity.
Persuasive Communication focuses on marketing and health communication messages, and investigate how companies, consumers, health providers, and clients communicate, and how these interactions influence their attitudes, cognition, and behaviour, with a strong focus on new media technologies.
Political Communication & Journalism focuses on understanding the role of media and communication in politics and democratic societies. The research is focused on how political news and information is produced, what the features of the contents are, and what the dissemination patterns, uses, and effects are.
Youth & Media Entertainment focuses on the role of entertainment media in the lives of young people. At its core, research in this group relies on a media psychology paradigm to answer questions about how different users are affected by different media content in different ways.

Cross-fertilisation and acceleration of research via research priority areas (RPAs)

Cross-fertilisation between ASCoR's lines of research has been stimulated by the university's policy of (competitive) multidisciplinary research priority areas (RPAs) since 2010. At the level of the Faculty of Social and Behavioural Sciences (FMG), as well as the level of the university. In 2010, the FMG decided to fund the RPA Entertainization of Society programme with Patti M. Valkenburg as principal investigator. The programme focused on the intended and unintended effects of entertainisation, which refers to the inclusion of entertainment-related elements, such as emotions, dramatic conflict, and sensational content, in information, education, commercials, and health-education messages. This interdisciplinary RPA included several

ASCoR researchers and colleagues from political science, informatics, psychology, and the medical sciences. One year later, the RPA refocused its programme under the title "Communication" and, now directed by Claes H. de Vreese and Patti M. Valkenburg, investigated the contents, uses, and consequences of media and communication more broadly, pioneering in a renewal of media effects theories.

The RPA Communication was renewed in 2018 by FMG after a positive evaluation of the first years. In this new stage, the RPA launched the Digital Communication Methods Lab (digicomlab), directed by Claes H. de Vreese, Theo Araujo, and Judith Möller. Between 2018 and 2022, the RPA focused on digital communication methods, ranging from mobile communication to artificial intelligence, both as methods and as objects of study. The RPA was successfully completed in 2022 and a plan was approved by ASCoR and the FMG to continue with the Digital Communication Methods Lab as part of ASCoR until at least 2027.

ASCoR researchers continued having a central role in the leadership of other FMG RPAs. The RPA Youth Digitality, launched in 2022, is co-directed by ASCoR's Jessica Taylor Piotrowski and Eline Smit. This RPA focuses on how young people are shaped by the digital world and how they navigate the intricate connections between the offline and online world, digital platforms, and their interactions with society and others in the digital context. Bert Bakker and Christin Scholz have been involved with the launch and coordination of the RPA Polarisation (also launched in 2022), which fosters interdisciplinary examination of causes, consequences, mechanisms, and potential interventions targeting (de)polarisation at the cognitive/individual, social/group, and societal levels. In 2023, the Faculty of Social and Behavioural Sciences launched three new RPAs: Artificial Intelligence & Politics (with ASCoR's Theo Araujo, Hilde Voorveld, and Claes H. de Vreese), Conflict & Society (with ASCoR's Alessandro Nai), and Bridges, bridging social and behavioural science disciplines to galvanizes sustainability (with ASCoR's Marijn Meijers and Anke Wonneberger). In addition, several ASCoR members have participated as members of the FMG RPAs during this period.

The success of the RPA format and the interdisciplinary nature of ASCoR's research went beyond the level of the Faculty of Social and Behavioural Sciences. As of 2015, the University of Amsterdam provided funding for the project Personalized Communication, a collaboration between ASCoR's RPA and the university's Institute for Information Law (IViR). Directed by Claes H. de Vreese and Natali Helberger, this programme focused on empirical and normative research on the uses, effects, and implications of personalised communication in the areas of politics, health, and commerce. The Personalized Communication project was broadened after a successful evaluation in 2019, and it was renamed Information, Communication, and the Data Society. The focus became the way artificial intelligence (AI) and algorithms affect the role, impact, and regulation of information and communication in our data society.

ASCoR researchers also have been active members and held leadership roles in other interfaculty RPAs. For example, the RPA Urban Mental Health, a collaboration between FMG researchers and the Faculty of Medicine (A-UMC) and the Faculty of Science (FNWI), is co-directed by ASCoR's Julia van Weert. This RPA focuses on the understanding of the dynamics of mental health problems in an urban environment with special attention to common mental health problems: depression, anxiety, and addiction. The RPA Human(e) AI, a collaboration between FMG, Amsterdam Law School, FNWI, and the Faculty of Humanities (FGw), stimulates new research at UvA on the societal consequences of the rapid development of AI and automated decision-making (ADM) in a wide variety of societal areas, being co-directed by Claes H. de Vreese. The RPA Trust in the Digital Society, a collaboration of FMG, Amsterdam Law School, the Faculty of Economics and Business (FEB), and FGw, focuses on the emerging distrust within our societies and on interventions that re-establish trust and trustworthiness at an interpersonal and institutional level, is co-directed by Theo Araujo. The wide involvement of ASCoR members in all RPAs demonstrates the deep embedding and contributions of ASCoR within the University of Amsterdam.

These multidisciplinary RPA programmes had a major effect on the collaboration of researchers within ASCoR and on collaboration between ASCoR researchers and the wider UvA community. The increase in collaboration within ASCoR can be seen in the growth of the average number of ASCoR authors per publication, increasing from approximately 1.2 in the early 2000s to almost 2 per publication in the early 2020s. An even more notable rise is the collaboration of ASCoR researchers with the wider community—within and outside of UvA—as evidenced by the average number of authors per publication of which an ASCoR member was part, growing from less than 2 in the early 2000s to almost 4 authors in the early 2020s.

Societal impact

To have a societal impact has always been one of ASCoR's aims, reflected both in its mission-oriented research agenda linked to crucial societal developments, and in its connection with societal partners, with whom ASCoR members often conduct research and deploy interventions in practice. These partners include the government, policy bodies, NGOs, healthcare organisations, interest groups, media organisations, and companies, among many others. ASCoR researchers often engage in societally relevant discussions by issuing policy advice and public reports, by presentations and publications for societal partners, in their media appearances, and in their active contributions as members of several advisory boards of (public) institutions.

ASCoR's societal engagement history is too long and varied for an exhaustive list. Some notable examples in the past five years include the activities of its members during the COVID-19 pandemic, issuing advice and reports for policymakers and public health professionals,[4] or the institute's partnership with a leading national newspaper and polling agency to research the Dutch 2021 parliamentary elections.[5] They also include a variety of public reports on key societal challenges, such as public perceptions about artificial intelligence[6] or social media usage, and the well-being of adolescents.[7] They also encompass interventions with public institutions or societal partners, such as an initiative to measure and improve the digital competence of Dutch citizens[8] in collaboration with the Ministry of the Interior and Kingdom Relations. Moreover, ASCoR members also provided research-based (policy) advice on multiple topics, such as on the disclosure of online sponsored content for minors in collaboration with the Dutch Media Authority, to the Trimbos Institute on drugs and addiction, to the Ministry of Health, Welfare and Sport on the development of health interventions, or to policymakers on micro-targeting. ASCoR members are also members of several (advisory) boards of key societal or public organisations, including of the European Digital Media Observatory and its Dutch–Flemish counterpart,[9] a collective of fact-checking organisations, media literacy experts, and academic researchers focusing on disinformation, the Complaints Council for the Netherlands Institute for the Classification of Audiovisual Material (NICAM), or of the Netherlands Gambling Authority, among many others.

Important vehicles for ASCoR's societal impact objectives are its affiliated research centres that collaborate with societal actors and disseminate

4 https://www.healthcommunication.nl/blog/category/covid_19.

5 https://www.ioresearch.nl/actueel/digitale-campagne-politieke-partijen-onder-de-loep/.

6 https://www.digicomlab.eu/reports/2018_adm_by_ai.

7 https://www.awesomescience.nl/publieksrapport.

8 https://www.dedigiq.nl.

9 https://edmo.eu/ and https://benedmo.eu/, respectively.

scientific knowledge. They include SWOCC (Foundation for Scientific Research on Commercial Communication, founded by the highly respected advertising practitioner and professor Giep Franzen),[10] CPC (Center for Politics and Communication),[11] CcaM (Center for Research on Children, Adolescents, and the Media),[12] and ACHC (Amsterdam Center for Health Communication).[13] In addition, ASCoR has endowed chairs that provide a bridge to society. These include the Chair in Strategic Communication, established by Logeion, the platform for communication professionals, the Chair in Customer Media, established by the Customer Media Chair Foundation, and the Chair in Information Society, established by the Royal National Library of the Netherlands.

ASCoR's PhD programme

Today ASCoR's PhD programme follows the department's PhD programme that started in 1993 with the first generation of PhD students. The pre-ASCoR generation of students were supervised by Denis McQuail and Peter Neijens and included Simone Bergman, Lilian van der Bolt, Patrick Hendriks, Bart van den Hooff, Andra Leurdijk, John Nouwens, Edith Smit, Marit Vochteloo, and Mir Wermuth.

The main aim of ASCoR's PhD programme is to promote the training of young academics in the study of fundamental scientific research issues in the field of communication science. Students are trained to extend, deepen, and actively apply their knowledge of theories and methodologies in order to become fully qualified, independent researchers. The programme prepares students for careers in academia or to take up research, consulting, or management positions in market and opinion research organisations.

To achieve these goals, students are offered a combination of research, coursework, and teaching tasks. Research, including data collection, is conducted throughout the project. The project proposal is usually written by senior ASCoR faculty and fits into the ASCoR research programme. A PhD project consists of three or more studies, typically leading to peer-reviewed articles. PhD students are usually required to teach in the bachelor's programme so that they can develop into full-fledged scholars. The final

10 https://www.swocc.nl.
11 https://www.polcomm.org.
12 https://www.ccam-ascor.nl.
13 https://www.healthcommunication.nl.

project months are devoted to finishing the thesis, based on the articles written throughout the project. After approval of the thesis by a committee of experts, there is a public defence.

ASCoR was fortunate to have a very talented group of PhD students in its first cohort, including PhD students who later became full professors, including Moniek Buijzen, Mark Deuze, Jochen Peter, and Claes H. de Vreese (Table 1.3 shows the alumni who have become full professors). The programme has grown strongly over the years, and it currently has 35 students from countries including Germany, the UK, India, China, Sudan, and the Netherlands. In total, ASCoR's PhD programme has led to around 200 doctorate degrees. Alumni of the programme continued their careers in academia, in research institutes, and outside the university.

Table 1.3. ASCoR's PhD alumni who became full professors (graduation year in brackets)

Marjolein Antheunis (2009)—Tilburg University
Theo Araujo (2015)—ASCoR, University of Amsterdam
Hajo Boomgaarden (2007)—University of Vienna
Moniek Buijzen (2003)—Radboud University and then Erasmus University Rotterdam
Yael de Haan (2012)—University of Groningen
Claes H. de Vreese (2003)—ASCoR, University of Amsterdam
Mark Deuze (2002)—Faculty of Humanities, University of Amsterdam
Bart van den Hooff (1997)—Free University Amsterdam
Martin Kroh (2003)—Bielefeld University
Sanne Kruikemeier (2014)—Wageningen University
Sophie Lecheler (2010)—University of Vienna
Judith Moller (2013)—University of Hamburg
Jochen Peter (2003)—ASCoR, University of Amsterdam
Stijn Reijnders (2006)—Erasmus University Rotterdam
Esther Rozendaal (2011)—Erasmus University Rotterdam
Edith Smit (1999)—ASCoR, University of Amsterdam
Roderick Swaab (2005)—INSEAD, Paris
Damian Trilling (2013)—Vrije Universiteit Amsterdam
Caroline Wagner (2004)—Ohio State University
Tamara Witschge (2007)—University of Groningen

What contributed to ASCoR's trajectory?

In this section, we reflect on the factors that may have contributed to ASCoR's trajectory. Subsequently, we discuss ASCoR's orientation on media technological and societal changes, including its international, funding, output, and research methods orientation, as well as individual factors.

Media technological and societal changes

The many exciting developments in the media and communication landscape and their challenges and opportunities for individuals, organisations, and society gave rise to fundamental research questions that became part of ASCoR's agenda. These developments ranged from the liberalisation of the telecom markets and the arrival of commercial broadcasters in ASCoR's early years, via the rise of digital, online, social, and mobile media, to the widespread application of big data and AI in media and communication. Media and communication became omnipresent, affecting all aspects of life, including politics, economics, health, and well-being. This continues to make ASCoR's research extremely relevant for policymaking, organisations, and individuals (Neijens & Valkenburg, 2024, this volume).

International orientation

From the outset, communication science in the Netherlands was internationally oriented. The founder of the discipline in the Netherlands, Kurt Baschwitz, became a driving force behind various international initiatives, including the International Association for Media and Communication Research (IAMCR), currently one the biggest international communication associations worldwide. He also founded the English-language journal *Gazette* (currently *International Communication Gazette*) in 1955 (van Ginneken, 2018; Wieten, 2005).

The international orientation of communication science at the University of Amsterdam was significantly strengthened further with the appointment of Denis McQuail as full professor of communication science in 1977. McQuail had relocated from the University of Leeds in the UK, where his earlier research about the influence of television on election outcomes had already drawn broad attention from colleagues in Europe and the United States. As a result of McQuail's tenure in Amsterdam, the department extended collaborations with colleagues across Europe and North America during this period. McQuail is well-known for his book *Mass Communication Theory*, which has sold over 150,000 copies worldwide and has been translated in 22 languages so far. ASCoR alumnus Mark Deuze now carries the legacy of McQuail, updating and publishing new editions of the textbook.

ASCoR had an international and specifically Anglo-Saxon focus since its creation. The aim of ASCoR has always been to present its research at international conferences organised by IAMCR, the World Association for Public Opinion Research (WAPOR), and the International Communication

Association (ICA), the largest communication association worldwide), as well as to publish in the leading English-language academic journals in the field. The international orientation of ASCoR was also reflected in the appointments of professors Holli Semetko (from the USA) and Klaus Schönbach (Germany) in the early years of ASCoR.

The 1990s were also marked by a strong belief that internationalisation was key when the world was globalising at a rapid pace, particularly for a relatively small country such as the Netherlands. At that time, the Dutch government and academic funding organisations had already begun to invest heavily in the internationalisation of Dutch academia in order to make it a focal point of recognition and funding of academic research. With increasing globalisation and the rise of the internet, it had also become apparent that English would be the international leading language for the next decades to come, and ASCoR embraced this trend.

The international orientation of ASCoR is also shaped by ASCoR's International Advisory Board, which since 2015 has consisted of leading experts in the field. Currently its members are Amy Jordan (Rutgers University, USA), Frank Esser (University of Zurich, Switzerland), Robin Nabi (University of California, Santa Barbara, USA), Mette Morsing (Copenhagen Business School, Denmark), Dhavan Shah (University of Wisconsin-Madison), and Michael Slater (The Ohio State University, USA).

The international orientation of ASCoR did not mean that ASCoR does not also play an important role for communication science in the Netherlands. This is evident from its leadership of the national research school, the Netherlands School of Communications Research (NeSCoR), and its important contributions to the national scientific journal *Communication Science* and to the Netherlands–Flanders Communication Association (NeFCA).

Funding orientation

Funding opportunities are important policy-steering instruments in academia. In the early years, communication science at the University of Amsterdam lacked sufficient funding, while student numbers increased massively. As a result of the imbalance between funding and student numbers, the Dutch Inspectorate of Education highlighted, at the end of the millennium, the need to increase and stabilise the financial support for the communication science programme, ensuring its long-term sustainability. The University of Amsterdam eventually followed these recommendations.

In the late 1990s, the University of Amsterdam wanted to strengthen its research and education capacity by becoming a conglomerate of strong research and education institutes instead of a centralised university. Teaching institutes and research institutes were, therefore, given more autonomy, including more control over the acquisition and spending of budgets. The founding of the communication science research institute around this time thus meant more control by the discipline over research funding.

Despite assuming an improved financial position, funding for communication science research in the new research institute was not secure. In the 2000s, the Dutch government had already begun withdrawing funds from universities and transferring increasingly higher shares of these funds to the Dutch Research Council (NWO). The government's goal for Dutch academic research was to become more competitive and to secure the country a forefront position internationally. The NWO was entrusted with the task of selecting the most successful researchers and projects for funding. For ASCoR, this was another arena where it could gain more sustained recognition and funding.

Output

Having a primarily Anglo-Saxon orientation, ASCoR was aware that specific forms of academic research output were increasingly becoming key indicators of success. This trend had started earlier at American universities, where publications in academic research journals were an important criterion for research funding and job promotions, and PhD theses were typically completed within four years after the start of a doctoral programme. The same criteria were increasingly used by the UvA and the NWO to define research success and allocate funding.

In line with the new indicators of research success, the allocation of research time within ASCoR became tied to research publications. This policy increased the chances of success in obtaining competitive grants from the NWO, and soon after ASCoR's foundation the first grants were allocated to communication science at the UvA. PhD research projects at ASCoR were conceptualised by senior faculty and included a series of studies and papers to be written for academic research journals. This ensured that PhD research projects were highly structured and usually finished within the allocated time. Moreover, the projects regularly resulted in several publications in (leading) international journals, giving candidates strong starting positions for further academic careers. Overall, ASCoR's policy thus ensured that key indicators for research success of UvA as well as the NWO were increasingly met. This led to additional recognition and funding, which in turn contributed again to possibilities for further high-quality research and publications.

The integrating role of research methods

ASCoR's strong empirical orientation was highlighted in its foundational mission statement, which stated: "ASCoR's research is rooted in the empirical tradition and characterised by methodological pluralism. Research is aimed at theory development and testing based on data." These aims have materialised in ASCoR researchers' strong commitment to methodological rigour as well as in methodological discussions and innovations that often fulfil an integrating role across ASCoR's programme groups.

ASCoR research has been known for its focus on social-scientific research methods, especially within the quantitative tradition, including content analysis, experimental designs, surveys, longitudinal panels, diary studies, experience sampling methods (ESMs), and digital trace data donation. In addition, the community of scholars at ASCoR has embraced a qualitative research tradition, often applying it for theory development within new contexts—ranging from intercultural patient–doctor communication to better understanding of how individuals conceptualise communication with machines.

Throughout the past 25 years, ASCoR researchers have also consistently taken up new analytical methods to answer crucial research questions in an increasingly complex communication environment. Examples of this methodological orientation can be seen in the adoption of analytical methods such as structural equation modelling, mediation, and time series analysis, and, recently, endeavours in computational methods (Trilling et al., 2024) or the usage of idiographic methods of analysis ($N = 1$) for theorising person-specific media effects (Valkenburg et al., 2024).

ASCoR's methodological orientation is reflected in the formation of the Digital Communication Methods Lab—originally as part of the RPA Communication and now an integral part of ASCoR—which focuses on developing and applying a range of innovative digital and computational methods to communication research. The methodological orientation also includes the commitment to open source research software development, supporting, for example, infrastructures for content analysis, experiments with conversational agents, or data donation pipelines (Trilling et al., 2024, this volume). In addition, ASCoR aims to support, broaden, and institutionalise Open Science and FAIR data principles, not only in its research, but also in its PhD training programme and in the methodological support of the teaching programmes of the Communication Science department, notably the research master's programme at the Graduate School of Communication.

Individual factors

In addition to the structural factors that have influenced the development of ASCoR, it is the researchers who have made ASCoR what it is today. This includes all ASCoR researchers, PhD students, and support staff. With their commitment and talent, they acquired the grants, the insights, as well as the publications, and they contributed to the development of the discipline. Some examples: Former and current ASCoR members include a recipient of the Spinoza Prize (also known as the "Dutch Nobel Prize," the highest Dutch scientific award for researchers who belong to the absolute top of science), as well as recipients of European Research Council (ERC) grants, NWO Veni, Vidi, Vici grants, and NWO Gravitation grants. Former and current ASCoR members have been elected president of the ICA, been named a fellow of the ICA, became ICA division chairs, served as presidents of the European Advertising Academy, held international leading positions (such as editors-in-chief or associate editors of flagship journals in our discipline), received the Career Award from NeFCA, been chosen as members of KNAW, been named as chair of the De Jonge Akademie (The Young Academy), been appointed distinguished university professors, held the position of chair of the SSH Council of the Netherlands, received best paper awards at conferences, and served as members of government and industry bodies.

To conclude

In its first 25 years, ASCoR has met the goals it outlined in 1997, when its 35 members articulated ASCoR's first research programme. It has continuously produced high-quality research "rooted in the empirical tradition and characterised by methodological pluralism," with a focus on communication's social potential and effects. It has become an international organisation, with a vibrant PhD programme and a large community of researchers with a broad set of research topics and numerous collaborations with universities in the Netherlands and across the globe.

The institute's development across these 25 years has been both a reflection of the time and of the academic environment in the Netherlands and internationally. Some of the factors and policies that have led to its productivity and international orientation also have triggered discussions—not only in ASCoR, but also in other research institutes and research policy bodies in the Netherlands—about the risks of focusing on success indicators such as the number of publications or acquired funding, a highly performance-oriented culture, and publication pressure.

Throughout the past 25 years, ASCoR has often adjusted its priorities with an eye towards the future of its faculty and with a strong dedication to fulfil its academic and societal mission. While present since the institute's foundation, the emphasis on collaborations and teamwork has increased throughout the years, with a renewed focus on team science and the deployment of large-scale collaborative thematic data collection with participation of many ASCoR members. This emphasis on teamwork is also reflected in the strong collaboration between ASCoR and the teaching institutes—namely the College of Communication and the Graduate School of Communication—as evidenced by the launch of large-scale research and teaching departmental initiatives, such as Communication in the Digital Society, and Mis-/Disinformation.

The institute, now with over 90 members and a PhD programme with about 35 candidates, continues to move forward and to navigate a dynamic and complex (inter)national environment. Building on its foundations, the institute continues its emphasis on interdisciplinary collaboration and team science, on its methodological strengths, on creating lasting societal impact, and on increasing the diversity of its perspectives and of its international collaborations—all the while continuously looking for ways to create room for everyone's talent in line with the national discussions on Recognition & Rewards.[14]

By now, communication has been recognised as a central topic in societal debates and as key component in addressing crucial societal challenges and crises. One cannot think, for example, about solving the societal challenges articulated in the University of Amsterdam strategic plan's pillars—Responsible Digital Transformations, Resilient and Fair Society, Healthy Future and Sustainable Prosperity—without considering the role of media and communication. The ASCoR community continues to rise to the challenge by expanding and accelerating its research aimed at addressing these crucial societal issues.

References

Hemels, J. (2017). *Recognition and development of communication science in the Netherlands*. Amsterdam School of Communication Research, University of Amsterdam. https://ascor.uva.nl/about-ascor/history/history.html#The-History-of-Communication-Science-in-The-Netherlands

14 https://recognitionrewards.nl.

Neijens, P., & Valkenburg, P. M. (2024). How technological and societal developments shaped the agenda of ASCoR. In T. Araujo & P. Neijens (Eds.), *Communication research into the digital society: Fundamental insights from the Amsterdam School of Communication Research* (pp. 33–50). Amsterdam University Press. https://doi.org/10.5117/9789048560592_ch02

Research Review. (2020). *Research review communication science 2014–2019* [Report by an international committee that conducted a review in the context of the Standard Evaluation Protocol for Public Research Organisations by the Association of Universities in the Netherlands (VSNU), the Dutch Research Council (NWO), and the Royal Netherlands Academy of Arts and Sciences (KNAW)]. De Onderzoekerij.

Trilling, D., Araujo, T., Kroon, A., Möller, A. M., Strycharz, J., & Vermeer, S. (2024). Computational communication science in a digital society. In T. Araujo & P. Neijens (Eds.), *Communication research into the digital society: Fundamental insights from the Amsterdam School of Communication Research* (pp. 247–263). Amsterdam University Press. https://doi.org/10.5117/9789048560592_ch15

Valkenburg, P. M., Beyens, I., Bij de Vaate, N., Janssen, L., & van der Wal, A. (2024). Person-specific media effects. In T. Araujo & P. Neijens (Eds.), *Communication research into the digital society: Fundamental insights from the Amsterdam School of Communication Research* (pp. 233–245). Amsterdam University Press. https://doi.org/10.5117/9789048560592_ch14

van Cuilenburg, J., & Zwier, S. (Eds.) (1998). *Proceedings of the ASCoR opening conference 1997*. Het Spinhuis.

van Ginneken, J. (2018). *Kurt Baschwitz: A pioneer of communication studies and social psychology*. Amsterdam University Press. https://doi.org/10.2307/j.ctt1zxxxgf.1

Wieten, J. (2005). Kurt Baschwitz and the founding of *Gazette*. *Gazette*, *67*(6), 523–530. https://doi.org/10.1177/0016549205057542

About the authors

Peter Neijens is an Honorary Fellow and Emeritus Professor at ASCoR. He held the Chair in Media and Persuasion at the University of Amsterdam until 2019. His research focuses on persuasive communication, in particular media use and effects in advertising and public opinion. He was co-founder and scientific director of ASCoR from 1998 to 2005 and of the Graduate School of Communication from 2009 to 2013. Email: P.C.Neijens@uva.nl

Sandra Zwier is Senior Lecturer at the Department of Communication Science. She was the Research Manager of ASCoR during its foundational decade (1997–2007). E-mail: S.M.Zwier@uva.nl

Claes H. de Vreese is Distinguished University Professor of AI and Society with a special emphasis on media and democracy. He co-directs the Gravitation Programme AlgoSoc and the AI, Media and Democracy Lab. He holds the Chair in Political Communication and was Scientific Director of ASCoR from 2005 to 2013. E-mail: C.H.deVreese@uva.nl

Jochen Peter is a Full Professor and holds the Chair in Media Entertainment at ASCoR. He is Programme Group Director of Youth & Media Entertainment at ASCoR. He was Scientific Director of ASCoR from 2013 to 2017. His research centres on emerging technologies and the psycho-social development of young people. E-mail: J.Peter@uva.nl

Rens Vliegenthart is a Full Professor and holds the Chair in Strategic Communication at Wageningen University & Research. Previously he held the Chair in Media, Organisations, and Society and, from 2018 to 2022, was the Scientific Director of ASCoR. His research focuses on media content and effects, both on citizens and public opinion, as well as on politicians and political decision-making. E-mail: Rens.Vliegenthart@wur.nl

Theo Araujo is a Full Professor of Media, Organisations and Society, and Scientific Director of ASCoR as of 2022. His research investigates the dynamic interplay between media and organisations, and what it means for society, with a special focus on trust and technology. His research has also a methodological focus on computational communication science and the implementation of large-scale data collection and analysis for communication research. E-mail: T.B.Araujo@uva.nl

Appendix 1.1. Full professors (chair holders) in communication science and their fields of study (*leeropdracht*) at the UvA appointed in the pre-ASCoR period

Mass communication pillar

Kurt Baschwitz (1947–1956)
  Theory of the press, public opinion, and propaganda; Press science and mass psychology
Vacancy (1956–1958)
Maarten Rooij (1958–1972)
  Theory of communication media, especially the press
Vacancy (1972–1977)
Denis McQuail (1977–1997)
  Mass communication

Mass psychology pillar

Marten Brouwer (1971–1994)
  Psychology of collective behaviour; Political psychology
Holli Semetko (1995)
  Audience and public opinion research

Public information systems pillar

Jan van Cuilenburg (1988)
  Informational aspects of public information systems; Communication policy

Other chairs

Cees Hamelink (1983)
  International communication
Joan Hemels (1986)
  History of the press, propaganda, and public opinion; Communication history

Note: Within brackets: date of appointment as full professor. An end date is included if the person had left before ASCoR was launched.
Source: https://albumacademicum.uva.nl

Appendix 1.2. ASCoR's management and staff from 1997

Directors
Jan van Cuilenburg (1997)
Peter Neijens (1998–2005)
Claes H. de Vreese (2005–2013)
Jochen Peter (2013–2017)
Rens Vliegenthart (2018–2022)
Theo Araujo (2022–)

Research managers
Sandra Zwier (1997–2007)
Maaike Prangsma (2008–2015)
Bas Sietses (2014–2017)
Jasper van de Pol (2018)
Amber van der Wal (2019–)

Office support
Margriet Smit
Pieke Assmann
Maaike Dudink
Kathleen Hair
Irene Lannoye
Willemijn van Maanen
Esther Nipperus
M. Pitambersingh
Miriam van der Putte
Hester Riethof
Ania Tekien
Berry Vermolen

Policy advice
Tijmen Lansdaal

Data stewards
Aino Koho
Monica Lodi

Appendix 1.3. Research programmes and programme directors from 1997

Until 2000

Communication, organisation, and policies (Jan van Cuilenburg, Jo Bardoel, Jan de Ridder, Betteke van Ruler)
Communication effects (Holli Semetko, Cees van der Eijk, Peter Neijens)
Media, audiences, and culture (Klaus Schönbach)

2001–2007

Media, journalism, and public opinion (Klaus Schönbach)
Media entertainment and popular culture (Patti M. Valkenburg, Liesbet van Zoonen)
Persuasive communication (Peter Neijens)

2008–2013

Corporate communication (Rens Vliegenthart, programme launched in 2013)
Persuasive communication (Edith Smit, Julia van Weert)
Political communication and journalism (Klaus Schönbach, Claes H. de Vreese)
Youth and media entertainment (Patti M. Valkenburg, Hans Beentjes)

2014–2019

Corporate communication (Martine van Selm, Rens Vliegenthart, Piet Verhoeven, Claartje ter Hoeven, Toni van der Meer)
Persuasive communication (Guda van Noort)
Political communication and journalism (Claes H. de Vreese)
Youth and media entertainment (Jessica Taylor Piotrowski, Jochen Peter)

From 2019

Corporate communication (Christian Burgers, Toni van der Meer, Piet Verhoeven)
Persuasive communication (Marieke Fransen, Guda van Noort, Eva van Reijmersdal, Hilde Voorveld)
Political communication and journalism (Linda Bos, Claes H. de Vreese)
Youth and media entertainment (Jochen Peter)

2. How Technological and Societal Developments Shaped the Agenda of ASCoR

Peter Neijens and Patti M. Valkenburg

Abstract

This chapter highlights characteristic lines of research in the 25-year history of the Amsterdam School of Communication Research (ASCoR) against the backdrop of key technological launches and changes in the media and communication landscape. We roughly distinguish four subsequent eras over the past 25 years. Each of the four eras, with their unique media developments and issues, gave rise to fundamental research questions that became part of ASCoR's agenda. Our overview also shows the programmatic mission of ASCoR's research throughout its history: The development of theories that help us understand the rapidly changing digital society, along with a strong emphasis on empirical research to explore, confirm, or disconfirm these theories.

Keywords: media and communication scholarship, institutional development, discipline formation, media and communication research, media history

Introduction

Over the past 25 years, the topics on the research agenda of the Amsterdam School of Communication Research (ASCoR) ranged from political communication to entertainment, from health communication to (deceptive) advertising, and from public opinion to family relations, and included communicators such as journalists, social media influencers, robots, and chatbots. Since the

Neijens, P., & Valkenburg, P. M. (2014). How technological and societal developments shaped the agenda of ASCoR. In T. Araujo & P. Neijens (Eds.), *Communication research into the digital society: Fundamental insights from the Amsterdam School of Communication Research* (pp. 33-50). Amsterdam University Press. https://doi.org/10.5117/9789048560592_ch02

foundation of ASCoR in 1997, its mission has been to contribute to knowledge and insights about the way media and communication influence society, organisations, and individuals. In doing so, the institute takes a broad view of media and communication, including traditional media (e.g., television) and new technologies (e.g., social media, virtual assistants, and robots), mass communication, and (mediated) interpersonal communication.

On the occasion of its 25th anniversary, we reflect on the contributions ASCoR has made to the international scientific community and society. Communication science is unique in the sense that no other social science discipline has centred its expertise around technological changes and their impact on individuals and society at large. Our goal is to put the contributions in this book in the perspective of the history of ASCoR's research agenda, and to show how its research has evolved, and how that research has responded to key changes in media and communication.[1]

It is impossible to discuss all studies and contributions of ASCoR over the past 25 years, considering that researchers in ASCoR have published over 3,000 refereed articles, books, and book chapters in this period. What we have in mind for this chapter is to highlight several characteristic lines of research against the backdrop of key technological launches and changes in the media and communication landscape of the past 25 years. In the next section we present a list of these changes and we roughly distinguish four subsequent eras. In the sections that follow we discuss typical studies from each era, selecting examples from the different research groups. Some studies included in this chapter, which are exemplary for the period in question, were not published until later due to, among other things, lengthy review and publication processes. For this chapter we have taken great advantage of ASCoR's annual and self-assessment reports since 1997, prepared by ASCoR's directors and research managers. We used the information from these reports to inform our overview and adopted the verbatim descriptions of the research programme and mission whenever appropriate.

Technological developments in the digital society: Four eras

It is almost a truism: The past 25 years have witnessed numerous fundamental changes in the media and communication landscape. Many of these changes have been important impetuses of ASCoR's research agenda.

1 A more extensive and broader perspective on the recognition and development of communication science in the Netherlands can be found in Hemels (2017).

Table 2.1 shows key technological launches and media changes influencing ASCoR's research, from the liberalisation of the telecom markets in the pre-ASCoR period, and the start of the internet with relatively "simple" affordances in the early years, through the rise of digital, online, social, and mobile media, to the widespread application of big data and AI in media and technology.

Table 2.1. Key media and communication launches from 1989

1989	Liberalisation of the telecommunication markets; rise of commercial broadcasters
1993	Internet for the general public (in the Netherlands)
1997	ICQ
1999	MSN Messenger; Nu.nl; Epinions
2001	Smart phones (Ericsson, BlackBerry)
2002	Google Search (Dutch language)
2002	Google News
2003	Skype
2004	Facebook (USA)
2005	Hyves (Dutch social medium); YouTube; Google Review
2006	Twitter
2008	iPhone (in the Netherlands)
2008	Facebook (Netherlands); App Store
2009	WhatsApp
2009	News apps
2010	Instagram; iPad
2011	Facebook's Messenger; Snapchat
2012	Tinder; Google Play Store; NOS app (Netherlands)
2017	TikTok
2022	ChatGPT; DALL-E; Bard; Ernie

In this chapter, we roughly distinguish four subsequent eras in the development of media and communication technologies. First, the Deregulated Media Markets Era, in the Netherlands from about 1990, which was late compared to other countries in Europe (Bardoel & van Cuilenburg, 2003). This first era is characterised by the liberalisation of the telecommunication markets, the arrival of commercial broadcasters, and the modest start of the internet.

In the second era (from about 2002) the emergence of digital technologies stimulated developments such as online news, games and intranets. Particularly relevant for this era is that Facebook (2004), YouTube (2005) and other social media entered the market. We label this the Online and Social Media Era.

Starting in 2007, after the introduction of the iPhone, our screens progressively moved from our desktops to our pockets so that we became permanently connected online. In this third era digital media and communication became omnipresent, influencing all aspects of life. This era also shows a trend toward the personalisation of media content, tailored to our interests, predispositions, and needs. We call this the Mobile and Personalised Communication Era.

The final era that we distinguish is the Big Data and AI Era. This era started around 2015. From that year, media applications collect massive amounts of data of their users, and AI infiltrates these applications with algorithms, augmented and virtual reality, chatbots, and virtual assistants. Challenges and opportunities that this era offer include misinformation and manipulation, algorithmic bias and discrimination, privacy concerns, and the increasing importance of influencers.

The Deregulated Media Markets Era (from approx. 1990)

Until 1990, the telecommunications companies (mail and telephone) in the Netherlands were state owned, there was only public broadcasting, and the internet was not yet a significant factor. The liberalisation of the telecommunications market (in 1989), the rise of commercial broadcasters (in 1989), and the launch of early internet applications (about 1993) completely changed the media landscape, leading to dual markets (public and commercial channels), (increased) competition, advertising dependency, the rise of new "popular" media formats, and the fragmentation of audiences. These fast changes in media market structures triggered debates about the quality of the media on offer and the governance of media as public good. These questions and concerns became a vital source of inspiration for studies in the pre-ASCoR and the early ASCoR period.

A first set of research questions involved the impact of the new market structures on media performance characteristics such as openness, diversity, social responsibility, and accountability, and its implications for media policy. Based on general economic theories and longitudinal studies on television, newspapers, and professional information markets in different countries, Jan van Cuilenburg and Richard van der Wurff showed that moderate levels of media competition and concentration in media markets improved innovation and diversity, whereas ruinous competition produced excessive uniformity of media products (van der Wurff & van Cuilenburg, 2001). Which policies could contribute to media market performance was examined by Jo Bardoel and Leen d'Haenens (2004). They assessed (best) practices in Europe and argued for accountability mechanisms, such as an ombudsperson, types of self-regulation, and types of public consultation (e.g., civic journalism).

Another development in the changing media markets was the declining interest in traditional newspapers and the emergence of free and online newspapers. A team led by Klaus Schönbach conducted a large-scale study on the use, reception, and effects of traditional media compared to online media. Traditional media such as newspapers, magazines, television, and radio, offer pre-arranged information, often also contextualised and interpreted (i.e., "display media"). Online media invite its users to put together one's own individual news and information diet (i.e., "search media"), with possible consequences for the user's knowledge about public affairs. Results from their study showed that both display and search media can make readers aware of a wide range of public affairs—but only in specific segments of society. Whereas people with little interest in public affairs seemed to profit from printed newspapers, only the best-educated group of online readers showed a greater diversity of topics of interest (Schoenbach et al., 2005).

Piet Bakker's research focused on the new free dailies that were becoming popular at the time (e.g., *Metro*, *Spits*), which were distributed via supermarkets and train stations. Bakker (2007) showed that the consequences of the new free dailies for current affairs journalism were quite minimal, as the professional level of their journalists did not differ from that of journalists of traditional (subscription) newspapers. He also showed that the introduction of free newspapers had a downward effect on the revenues of traditional newspapers, as pressure on advertising markets had increased, damaging the financial bases of most incumbent publishers.

The arrival of commercial broadcasters on the market and the associated increased competition among media channels led to the emergence of popular media formats such as talk shows, soap operas, and infotainment programmes, which were studied by Liesbet van Zoonen and colleagues. They investigated how politicians, women, and minorities were represented in these programmes, and how these programmes contributed to the construction of meaning, identity, and citizenship in its users. It was shown, for instance, that the popular media formats facilitated political communication that produced highly personalised and schematic frames that are characterised by moral rather than political evaluations (van Zoonen, 1998). Kees Brants (1998) concluded on the basis of his research that the alleged dangers of the new infotainment format (e.g., little attention to political content; negative campaigning) were based on questionable premises.

The increasing importance of European integration for Europe-wide and national politics during this period led to studies on agenda setting, priming, and framing of European integration in the news and their impact on public opinion, European elections, and European Union (EU) referendums. Holli Semetko and her colleagues employed a multi-method approach, including interviews with journalists, newsroom observation, content analyses of media and party programmes, longitudinal surveys (in all EU countries), and experiments. It was shown that news frames have important effects on opinions about political and economic issues, on how Europeans think about European integration and the single currency, and how this relates to country characteristics (Semetko & Valkenburg, 2000; Semetko et al., 2000).

The Online and Social Media Era (from approx. 2002)

Around 2002, ASCoR started researching the social implications of "social media," a term that had not yet been widely accepted at the time. Although mobile phones already existed, social interaction was still mainly via desktops. In this period, ASCoR researchers massively started to uncover the ethical, individual, and societal consequences of "online communication," "computer-mediated communication" (CMC), "intranets" (private networks within organisations), and "information and communication technologies" (ICTs) in organisations and families and among citizens, patients, and peers.

In the organisational context, CMC started to play an increasingly important role. The intended and unintended effects of the introduction of new ICTs in organisations were studied by Jan de Ridder and Bart van den Hooff. Topics included the adoption of email in organisations, and the effects of ICTs on social networks and knowledge sharing within and between organisations. For example, Bart van den Hooff conducted several longitudinal case studies into the value of intranets for organisations. His studies showed that the primary value of intranets was in providing information, not in facilitating communication. Moreover, accessibility, user friendly navigation, and a good search engine proved crucial for establishing optimal use and effectiveness of intranets (van den Hooff & de Ridder, 2004).

Around the same period, a team led by Tom Postmes examined the effects of CMC on social processes in interpersonal and group communication within the framework of the social identity model of deindividuation effects (SIDE model). Their research showed that despite the "minimal" transmission of social information afforded by CMC devices, the social effects of information shared in such devices on group identity can be maximal because normative influences are more accentuated in online than in comparable offline groups (Postmes et al., 2001).

Other researchers investigated the cultural meanings of online communication. For example, Sally Wyatt investigated patients' internet use compared to the use of, and trust in, other sources of information, such as patient folders. Her results revealed that patients rarely rely on one single source of online information but draw on a variety of them (Henwood et al., 2003). Such an active adaptation of the internet and other new media to personal needs also emerged from studies by Liesbet van Zoonen about the uses of the internet in families. She found, for example, that family members often have their unique internet applications to inform and entertain themselves and to maintain their social capital (van Zoonen, 2002).

In 2003, Patti M. Valkenburg and Jochen Peter started to investigate the social consequences of social media focusing on platforms like Hyves, MSN Messenger, and CU2, none of which still exist today. Their publications led to a fundamental re-evaluation of previous, mainly USA-based, research, which at that period had primarily reported negative social consequences of online communication. Together with Marjolijn Antheunis, Alexander Schouten, Susanne Baumgartner, and Sindy Sumter, they showed, for example, that online communication triggers online inhibition, which in turn activates online (sexual) risk behaviour. But they also showed that online communication can result in higher quality friendships and romantic relationships because it induces intimate self-disclosures online, which, in turn, enhance friendship closeness and romantic relationship formation (Antheunis et al., 2007; Baumgartner et al., 2012; Schouten et al., 2007; Valkenburg & Peter, 2009).

The rising popularity of online gaming inspired several ASCoR researchers. For example, Jeroen Jansz and colleagues confirmed in a series of studies with various co-authors (e.g., Jansz & Martens, 2005; Jansz & Tanis, 2007) the stereotype of the gamer (young men who spend a lot of their leisure time on gaming), but also that games are not played in isolation. They showed that motivations with respect to competition, challenge, and social contact were important for gamers. Jeroen Lemmens, Patti M. Valkenburg, and Jochen Peter developed a scale for game addiction and showed that social competence, self-esteem, and loneliness were significant predictors of pathological gaming (Lemmens et al., 2009).

The new high-choice media landscape brought about another new phenomenon, which has been referred to as "brand placement," "sponsored content," or "deceptive advertising." These concepts all refer to the integration of persuasive messages into seemingly non-commercial content with the aim of influencing the media user. Eva van Reijmersdal, Peter Neijens, and Edith Smit showed that audience characteristics (e.g., attitudes), placement characteristics (e.g., prominence), and contextual factors (e.g., programme type) influence the impact of brand placement on brand knowledge, attitude, and image (van Reijmersdal et al., 2009).

The rapidly expanding commercial media environment called for a better understanding of young people's advertising literacy. Together with Esther Rozendaal, Eva van Reijmersdal, and Patti M. Valkenburg, Moniek Buijzen started a programmatic line of research to understand the potential negative effects of this environment, such as parent–child conflict and materialism (Buijzen & Valkenburg, 2003). Buijzen and colleagues also conducted several studies to better understand children's advertising literacy (e.g., Rozendaal et al., 2011). Finally, they introduced an influential framework to understand young people's cognitive and affective processing of commercial content (Buijzen et al., 2010).

Finally, Claes H. de Vreese, Holli Semetdo, and Hajo Boomgaarden continued their research on political communication in the expanding EU with a study of news coverage of the 2004 European parliamentary elections in all 25 member states. Their study provided a unique pan-European overview of the campaign coverage. They found that national political actors (and not EU actors) dominated the news of the elections. In addition, they found that the news in the 15 old member states was generally negative towards the EU, whereas in the 10 new countries a mixed pattern was found, meaning that the broadsheet press and television news were, on average, positive while the tabloid papers were, on average, negative towards the EU (de Vreese et al., 2006).

The Mobile and Personalised Communication Era (from approx. 2007)

This era is characterised by the development of the smartphone and mobile apps and the growth of mobile devices. In 2008, Apple's App store was launched with 500 mobile apps, followed by the Google Play Store in 2012. Although initially the most successful mobile apps were "freemium" games (e.g., *Angry Birds*), mobile apps soon started to cover entire lifestyles, including social networking (Facebook, 2008), dating (Tinder, 2012), news reading (NOS, 2012), book reading (Kindle, 2009), travelling (Uber, 2012), banking (ING, 2013), shopping (Bol, 2014), and even meditation (Calm, 2012). Other hallmarks of this era are the huge trend towards personalised media content, as well as the blurred distinctions between private and public content, between informative, entertaining, and persuasive content, and between professionally and non-professionally produced media content.

In this era, the same entertainment products (shows, films, video games) can be accessed through an endless stream of mobile channels and devices. This changing entertainment environment inspired Patti M. Valkenburg to investigate the "entertainisation of childhood." Together with Jessica Taylor Piotrowski, Ine Beyens, and other colleagues, Valkenburg found no evidence for the societal concern that the overwhelming abundance of fast-paced entertainment media would enhance ADHD-like behaviours (i.e., attention problems and impulsivity). On the other hand, they did find that playing video games could stimulated adolescents' level of problem-solving intelligence (Nikkelen et al., 2014; Fikkers et al., 2019). The results led Valkenburg and Peter to develop the differential susceptibility to media effects model (DSMM) (Valkenburg & Peter, 2013), a transactional media effects model to understand when, why, and who is more or less susceptible to media effects.

Online media offer ample opportunities to personalise or tailor online information to target groups. Ewa Maslowska, Edith Smit, and Bas van den Putte investigated how different personalisation strategies (in which personal cues are added to a message) influenced the effects of the advertising and found, for example, that perceived personalisation increased attention to the message, which stimulated both positive and negative thoughts about the message (Maslowska et al., 2016). Similar work in health communication emerged. Julia van Weert and her team showed that tailoring the type of information to the patient's personal needs was associated with fewer perceived barriers to medication intake. Tailoring the message to participants' learning styles and mode preferences increased their motivation to attend and process the information, facilitating learning and recall of information, as well as healthy behaviour change (Smit et al., 2015).

Addressing the role of online media in organisational communication, Ward van Zoonen, Joost Verhoeven, and Rens Vliegenthart focused on the use of Twitter by employees in mid- and large-size organisations. Based on an elaborate content analysis, they developed a typology of work-related Twitter use based on the distinction between the profession-related, organisation-related, and work-related content of tweets. They found, for example, that Twitter can enhance the integration of personal and professional life domains, because employees often tweet about their work outside regular work hours, but they also tweet on a personal title when being at their work (van Zoonen et al., 2016).

Another characteristic of this era is the increased agency of individuals to obtain and select online information, for example, via recommendation systems or reactions to online news or health information. Lotte Willemsen, Peter Neijens, Fred Bronner, and Jan de Ridder studied these new interactive processes within the realm of electronic word of mouth, the sharing of reviews or recommendations on social media or websites. They found that the number and diversity of arguments in the review and its valence (positive or negative) played an important role in its perceived usefulness by readers (Willemsen et al., 2011). Interestingly, Julia van Weert and Bas van den Putte and colleagues concluded that such agency to obtain online information does not benefit all individuals. Individuals with (1) a low health literacy, (2) little ability to evaluate online health information, and (3) little trust in the internet as a source of health information were less able to evaluate online health information than individuals who scored higher on these skills and perceptions (Diviani et al., 2015).

In this period, ASCoR continued in advancing framing effects theory and research, both by focusing on the theoretical underpinnings of the effects (e.g., de Vreese & Lecheler, 2012), the duration (e.g., Lecheler & de Vreese, 2013), and the conditional indirect effects (e.g., Schuck & de Vreese, 2012) showing, for instance, that framing effects on political mobilisation is moderated by prior attitudes and mediated by risk perceptions.

In the field of corporate communication, Toni van der Meer and colleagues studied the framing of organisational crisis situations by the "domains" of public relations (in press releases), the news media (in newspaper articles), and the public (on social media). They showed that the degree of "alignment" (congruence) in the frames differed over the course of a crisis. After the initial phase the domains interact and align their frames (possibly to reduce ambiguity and uncertainty), and in the final phase the frames de-align (move away from each other) according to the specific interests and goals of the domains (van der Meer et al., 2014).

Marieke Fransen and Claartje ter Hoeven (2013) examined how managers and executives can improve the effectiveness of their negative written communications (i.e., refusal of employees' requests) by incorporating the concept of "fit" into their message framing. They showed that a fit between the framing of the message (using promotion-related or prevention-related terms) and the employee (who focuses either on the attainment of positive outcomes, or on the avoidance of negative outcomes) leads to more favourable outcomes.

Finally, this era witnessed a growing populism in the Netherlands (and beyond). More so than ever, new populist parties started to depend on the media for their electoral breakthrough. This growing populism in the Netherlands was on the research agenda of Linda Bos, Wouter van der Brug, and Claes H. de Vreese. Indeed, they showed that more successful right-wing populist leaders (e.g., Rita Verdonk, Geert Wilders) were more prominent during the national election campaign in the Netherlands in 2006, and that the most successful right-wing populist leader, Geert Wilders, also appeared more authoritative in the news (Bos et al., 2010).

The Big Data and AI Era (from approx. 2015)

This era is characterised by massive datafication, a rapidly increasing application of algorithms and AI in media and communication devices, and a huge surge in new computational methods, including automated content analyses, machine learning, and machine vision (Trilling et al., 2024). The many new methods to collect fine-grained data on the individual and collective level also came with the proliferation of powerful methods of analysis, such as time series analysis (Damstra et al., 2021), latent growth models (van der Wal et al., 2022), and person-specific analysis (Valkenburg et al., 2021). In this final section, we highlight a few exemplary studies and projects that appeared in this era; other examples of studies and projects typical for this era are discussed in various chapters of this book.

New computational methods were used in a study by Anne Kroon, Damian Trilling, and Tamara Raats on ethnic stereotypes in the news. Using automated content analysis and unsupervised machine learning, they analysed more than 3 million news articles. They showed that, compared with ethnic in-groups, news content proved to implicitly associate ethnic out-groups relatively strongly with low-status and high-threat stereotypes. In addition, they showed that, across time, content about ethnic out-groups has become progressively negative and remote from factual integration outcomes (Kroon et al., 2021). In addition, using a sample of more than 5,300 tweets and about 154,000 retweets from global top brands, Theo Araujo, Peter Neijens, and Rens Vliegenthart demonstrated how different types of Twitter users (influencers, information brokers, and strong ties) spread tweets through their networks and in doing so build word of mouth for brands (Araujo et al., 2017).

ASCoR's research during this period included studies of users' perceptions of disembodied conversational agents such as virtual assistants or chatbots. For example, Araujo and colleagues studied the influence of the conversational agents' language, gender, name, and modality on users' perceptions. It was shown that text-based assistants were perceived as more human compared to voice-based assistants (Ischen et al., 2022). Virtual assistants with a human name were perceived as more persuasive than assistants without a human name (Voorveld & Araujo, 2020), and the warmth (high vs low) of a chatbot's language and the chatbot's assigned gender elicited stereotypes that affected the perceived trust, helpfulness, and competence of the chatbot (Bastiansen et al., 2022).

Robots were central in studies by Jochen Peter and his team. For example, their studies showed that children readily accepted robots and that acceptance was further affected by social, adaptive robot behaviour, children's sex and age, as well as frequency of the interaction (van Straten et al., 2020). Their research also suggests that hedonic considerations (the perception that using the robot is enjoyable and pleasant) and normative considerations (the perception that robot use is approved by the family) predicted children's intention to adopt a domestic social robot (de Jong et al., 2022).

A possible negative effect of AI-driven communication is when it accelerates the spread of misinformation, contributing to problems for media trust and democracy. Hameleers (2021) showed that misinformation and disinformation are seen as relatively credible and persuasive by the public, and that people do not differentiate clearly between fact-free and evidence-based misinformation in their rating of message credibility and issue agreement. Effects of disinformation in the form of deepfakes showed that deepfakes can have the intended harmful effects (Dobber et al., 2021).

Finally, ASCoR's studies into the perception of algorithms also showed that media users' lack understanding of algorithm-driven targeting (Smit et al., 2014). This lack of understanding was especially prevalent among women, the elderly, and the low-educated (Zarouali et al., 2022). Trying to empower media users, for example, by giving them information about the personalisation process, did not influence consumers' motivation to opt out of personalised ads (Strycharz et al., 2019).

To conclude

Our overview shows how the dynamic communication environment and crucial media challenges have inspired ASCoR's research over the past 25 years. Each of the four eras that we discussed—the Deregulated Media Markets Era, the Online and Social Media Era, the Mobile and Personalised Communication Era, and the Big Data and AI Era—with their unique media developments and issues, gave rise to fundamental research questions that became part of ASCoR's agenda. Some of the early issues seem outdated now, for instance, the introduction of email in organisations, or the introduction of online newspapers—in that time new and highly relevant and researched in influential studies.

Other of the earlier issues became less urgent, but came back: issues of media performance and public policy were highly relevant in the years after the liberalisation of the telecommunication markets and the rise of commercial broadcasters, became less important in a period of media abundance, but are back on the agenda in the current era of Big Tech dominance with issues such as fake news, populism, polarisation, and privacy. Other issues, such as CMC, underwent a fundamentally different meaning with the advent of AI-dominated communication technologies, including social media (TikTok, YouTube, Instagram reels), robots, and chatbots. In this chapter, we have shown how advances in media and technologies have inspired theory formation and the empirical research of communication scholars in general and those of ASCoR in particular.

Our overview also shows the fundamental characteristics of ASCoR's research throughout its 25-year history. The "aim of studies" within ASCoR has always been the development of theories that help us understand the rapidly changing digital society, along with a strong emphasis on empirical research to explore, confirm, or disconfirm these theories. ASCoR's research places a strong emphasis on the different ways people create, use, and experience their media environment. Depending on the interaction between individual dispositions and contextual factors, people may differ considerably in how they use particular media content, how they communicate, and how they experience the consequences of this content. The fascinating range of highly relevant issues—individual, organisational, and societal—emerging from the dynamic developments in media and communication, and their implications for the way we live, work, play, thrive, and connect with others, continue to be a great source of inspiration for ASCoR's researchers.

References

Antheunis, M. L., Schouten, A. P., Valkenburg, P. M., & Peter, J. (2007). Computer-mediated communication and interpersonal attraction: An experimental test of two explanatory hypotheses. *CyberPsychology & Behavior, 10*(6), 831–836. https://doi.org/10.1089/cpb.2007.9945

Araujo, T., Neijens, P. C., & Vliegenthart, R. (2017). Getting the word out on Twitter: The role of influentials, information brokers and strong ties in building word-of-mouth for brands. *International Journal of Advertising, 36*(3), 496–513. https://doi.org/10.1080/02650487.2016.1173765

Bakker, P. (2007). Free daily journalism—Anything new? *Journalistica, 2*(4), 22–32. https://doi.org/10.7146/journalistica.v2i4.1801

Bardoel, J., & d'Haenens, L. (2004). Media meet the citizen: Beyond market mechanisms and government regulations. *European Journal of Communication, 19*(2), 165–194. https://doi.org/10.1177/0267323104042909

Bardoel, J., & van Cuilenburg, J. (2003). *Communicatiebeleid en communicatiemarkt. Over beleid, economie en management voor de communicatiesector.* Otto Cramwinckel Uitgever.

Bastiansen, M. H. A., Kroon, A. C., & Araujo, T. B. (2022). Female chatbots are helpful, male chatbots are competent? The effects of gender and gendered language on human–machine communication. *Publizistik, 67*(4), 601–623. https://doi.org/10.1007/s11616-022-00762-8

Baumgartner, S. E., Sumter, S. R., Peter, J., & Valkenburg, P. M. (2012). Identifying teens at risk: Developmental pathways of online and offline sexual risk behavior. *Pediatrics, 130*(6), e1489–e1496. https://doi.org/10.1542/peds.2012-0842

Bos, L., van der Brug, W., & de Vreese, C. H. (2010). Media coverage of right-wing populist leaders. *Communications, 35*(2), 141–163. https://doi.org/10.1515/comm.2010.008

Brants, K. (1998). Who's afraid of infotainment? *European Journal of Communication, 13*(3), 315–335. https://doi.org/10.1177/0267323198013003002

Buijzen, M., & Valkenburg, P. M. (2003). The effects of television advertising on materialism, parent–child conflict, and unhappiness: A review of research. *Journal of Applied Developmental Psychology, 24*(4), 437–456. https://doi.org/10.1016/S0193-3973(03)00072-8

Buijzen, M., van Reijmersdal, E. A., & Owen, L. H. (2010). Introducing the PCMC model: An investigative framework for young people's processing of commercialized media content. *Communication Theory, 20*(4), 427–450. https://doi.org/10.1111/j.1468-2885.2010.01370.x

Damstra, A., Jacobs, L., Boukes, M., & Vliegenthart, R. (2021). The impact of immigration news on anti-immigrant party support: Unpacking agenda-setting and issue ownership effects over time. *Journal of Elections, Public Opinion and Parties, 31*(1), 97–118. https://doi.org/10.1080/17457289.2019.1607863

de Jong, C., Peter, J., Kühne, R., & Barco, A. (2022). Children's intention to adopt social robots: A model of its distal and proximal predictors. *International Journal of Social Robotics, 14*(4), 875–891. https://doi.org/10.1007/s12369-021-00835-0

de Vreese, C. H., Banducci, S. A., Semetko, H. A., & Boomgaarden, H. G. (2006). The news coverage of the 2004 European parliamentary election campaign in 25 countries. *European Union Politics, 7*(4), 477–504. https://doi.org/10.1177/1465116506069440

de Vreese, C. H., & Lecheler, S. (2012). News framing research: An overview and new developments. In H. A. Semetko & M. Scammell (Eds.), *The SAGE handbook of political communication* (pp. 292–306). Sage. https://doi.org/10.4135/9781446201015.n24

Diviani, N., van den Putte B., Giani S., & van Weert, J. C. (2015). Low health literacy and evaluation of online health information: A systematic review of the literature. *Journal of Medical Internet Research*, *17*(5), Article e112. https://doi.org/10.2196/jmir.4018

Dobber, T., Metoui, N., Trilling, D., Helberger, N., & de Vreese, C. (2021). Do (micro-targeted) deepfakes have real effects on political attitudes? *The International Journal of Press/Politics*, *26*(1), 69–91. https://doi.org/10.1177/1940161220944364

Fikkers, K. M., Piotrowski, J. T., & Valkenburg, P. M. (2019). Child's play? Assessing the bidirectional longitudinal relationship between gaming and intelligence in early childhood. *Journal of Communication*, *69*(2), 124–143. https://doi.org/10.1093/joc/jqz003

Fransen, M. L., & ter Hoeven, C. L. (2013). Matching the message: The role of regulatory fit in negative managerial communication. *Communication Research*, *40*(6), 818–837. https://doi.org/10.1177/0093650211427140

Hameleers, M. (2021). *Populist disinformation in fragmented information settings: Understanding the nature and persuasiveness of populist and post-factual communication*. Routledge. https://doi.org/10.4324/9781003194668

Hemels, J. (2017). *Recognition and development of communication science in the Netherlands*. https://ascor.uva.nl/about-ascor/history/history.html#The-History-of-Communication-Science-in-The-Netherlands

Henwood, F., Wyatt, S., Hart, A., & Smith, J. (2003). "Ignorance is bliss sometimes": Constraints on the emergence of the informed patient in the changing landscapes of health information. *Sociology of Health and Illness*, *25*(6), 589–607. https://doi.org/10.1111/1467-9566.00360

Ischen, C., Araujo, T. B., Voorveld, H. A. M., van Noort, G., & Smit, E. G. (2022). Is voice really persuasive? The influence of modality in virtual assistant interactions and two alternative explanations. *Internet Research*, *32*(7), 402–425. https://doi.org/10.1108/INTR-03-2022-0160

Jansz, J., & Martens, L. (2005). Gaming at a LAN event: The social context of playing video games. *New Media & Society*, *7*(3), 333–355. https://doi.org/10.1177/1461444805052280

Jansz, J., & Tanis, M. (2007). Appeal of playing online first person shooter games. *Cyber Psychology & Behavior*, *10*(1), 133–136. https://doi.org/10.1089/cpb.2006.9981

Kroon, A. C., Trilling, D., & Raats, T. (2021). Guilty by association: Using word embeddings to measure ethnic stereotypes in news coverage. *Journalism & Mass Communication Quarterly*, *98*(2), 451–477. https://doi.org/10.1177/1077699020932304

Lecheler, S. K., & de Vreese, C. H. (2013). What a difference a day makes? The effects of repetitive and competitive news framing over time. *Communication Research, 40*(2), 147–175. https://doi.org/10.1177/0093650212470

Lemmens, J. S., Valkenburg, P. M., & Peter, J. (2009). Development and validation of a game addiction scale for adolescents. *Media Psychology, 12*(1), 77–95. https://doi.org/10.1080/15213260802669458

Maslowska, E., Smit, E. G., & van den Putte, B. (2016). It is all in the name: A study of consumers' responses to personalized communication. *Journal of Interactive Advertising, 16*(1), 74–85. https://doi.org/10.1080/15252019.2016.1161568

Nikkelen, S. W. C., Valkenburg, P. M., Huizinga, M., & Bushman, B. J. (2014). Media use and ADHD-related behaviors in children and adolescents: A meta-analysis. *Developmental Psychology, 50*(9), 2228–2241. https://doi.org/10.1037/a0037318

Postmes, T., Spears, R., Sakhel, K., & de Groot, D. (2001). Social influence in computer-mediated communication: The effects of anonymity on group behavior. *Personality and Social Psychology Bulletin, 27*(10), 1243–1254. https://doi.org/10.1177/01461672012710001

Rozendaal, E., Buijzen, M., & Valkenburg, P. M. (2011). Children's understanding of advertisers' persuasive tactics. *International Journal of Advertising, 30*(2), 329–350. https://doi.org/10.2501/IJA-30-2-329-350

Schoenbach, K., de Waal, E., & Lauf, E. (2005). Online and print newspapers: Their impact on the extent of the perceived public agenda. *European Journal of Communication, 20*(2), 245–258. https://doi.org/10.1177/0267323105052300

Schouten, A. P., Valkenburg, P. M., & Peter, J. (2007). Precursors and underlying processes of adolescents' online self-disclosure: Developing and testing an "internet-attribute-perception" model. *Media Psychology, 10*(2), 292–314. https://doi.org/10.1080/15213260701375686

Schuck, A. R. T., & de Vreese, C. H. (2012). When good news is bad news: Explicating the moderated mediation dynamic behind the reversed mobilization effect. *Journal of Communication, 62*(1), 57–77. https://doi.org/10.1111/j.1460-2466.2011.01624.x

Semetko, H. A., de Vreese, C. H., & Peter, J. (2000). Europeanised politics—Europeanised media? *West European Politics, 23*(4), 121–141. https://doi.org/10.1080/01402380008425403

Semetko, H. A., & Valkenburg, P. M. (2000). Framing European politics: A content analysis of press and television news. *Journal of Communication, 50*(2), 93–109. https://doi.org/10.1111/j.1460-2466.2000.tb02843.x

Smit, E. G., van Noort, G., & Voorveld, H. A. M. (2014). Understanding online behavioural advertising: User knowledge, privacy concerns and online coping behaviour in Europe. *Computers in Human Behavior, 32*, 15–22. https://doi.org/10.1016/j.chb.2013.11.008

Smit, E. S., Linn, A. J., & van Weert, J. C. M. (2015). Taking online computer-tailoring forward: The potential of tailoring the message frame and delivery mode of online health behaviour change interventions. *The European Health Psychologist, 17*(1), 25–31. https://www.ehps.net/ehp/index.php/contents/article/view/762

Strycharz, J., van Noort, G., Smit, E., & Helberger, N. (2019). Protective behavior against personalized ads: Motivation to turn personalization off. *Cyberpsychology: Journal of Psychosocial Research on Cyberspace, 13*(2), Article 1. https://doi.org/10.5817/CP2019-2-1

Trilling, D., Araujo, T., Kroon, A., Möller, A. M., Strycharz, J., & Vermeer, S. (2024). Computational communication science in a digital society. In T. Araujo & P. Neijens (Eds.), *Communication research into the digital society: Fundamental insights from the Amsterdam School of Communication Research* (pp. 247–263). Amsterdam University Press. https://doi.org/10.5117/9789048560592_ch15

Valkenburg, P. M., Beyens, I., Pouwels, J. L., van Driel, I. I., & Keijsers, L. (2021). Social media and adolescents' self-esteem: Heading for a person-specific media effects paradigm. *Journal of Communication, 71*(1), 56–78. https://doi.org/10.1093/joc/jqaa/039

Valkenburg, P. M., & Peter, J. (2009). The effects of instant messaging on the quality of adolescents' existing friendships: A longitudinal study. *Journal of Communication, 59*(1), 79–97. https://doi.org/10.1111/j.1460-2466.2008.01405.x

Valkenburg, P. M., & Peter, J. (2013). The differential susceptibility to media effects model. *Journal of Communication, 63*(2), 221–243. https://doi.org/10.1111/jcom.12024

van den Hooff, B., & de Ridder, J. A. (2004). Knowledge sharing in context: The influence of organizational commitment, communication climate and CMC use on knowledge sharing. *Journal of Knowledge Management, 8*(6), 117–130. https://doi.org/10.1108/13673270410567675

van der Meer, T. G. L. A., Verhoeven, P., Beentjes, H., & Vliegenthart, R. (2014). When frames align: The interplay between PR, news media, and the public in times of crisis. *Public Relations Review, 40*(5), 751–761. https://doi.org/10.1016/j.pubrev.2014.07.008

van der Wal, A., Pouwels, J. L., Piotrowski, J. T., & Valkenburg, P. M. (2022). Just a joke? Adolescents' preferences for humor in media entertainment and real-life aggression. *Media Psychology, 25*(6), 797–813. https://doi.org/10.1080/15213269.2022.2080710

van der Wurff, R., & van Cuilenburg, J. (2001). Impact of moderate and ruinous competition on diversity: The Dutch television market. *The Journal of Media Economics, 14*(4), 213–229. https://doi.org/10.1207/S15327736ME1404_2

van Reijmersdal, E., Neijens, P. C., & Smit, E. G. (2009). A new branch of advertising: Reviewing factors that influence reactions to product placement. *Journal of Advertising Research, 49*(4), 429–449. https://doi.org/10.2501/S0021849909091065

van Straten, C. L., Peter, J., & Kühne, R. (2020). Child–robot relationship formation: A narrative review of empirical research. *International Journal of Social Robotics, 12*(2), 325–344. https://doi.org/10.1007/s12369-019-00569-0

van Zoonen, E. A. (1998). A day at the zoo: Politicians, pigs and popular culture. *Media, Culture and Society, 20*(2), 183–200. https://doi.org/10.1177/016344398020002002

van Zoonen, L. (2002). Gendering the internet: Claims, controversies and cultures. *European Journal of Communication, 17*(1), 5–23. https://doi.org/10.1177/0267323102017001006

van Zoonen, W., Verhoeven, J. W. M., & Vliegenthart, R. (2016). How employees use Twitter to talk about work: A typology of work-related tweets. *Computers in Human Behavior, 55*, 329–339. https://doi.org/10.1016/j.chb.2015.09.021

Voorveld, H. A. M., & Araujo, T. (2020). How social cues in virtual assistants influence concerns and persuasion: The role of voice and a human name. *Cyberpsychology, Behavior, and Social Networking, 23*(10), 689–696. https://doi.org/10.1089/cyber.2019.0205

Willemsen, L. M., Neijens, P. C., Bronner, F., & de Ridder, J. A. (2011). "Highly Recommended!": The content characteristics and perceived usefulness of online consumer reviews. *Journal of Computer-Mediated Communication, 17*(1), 19–38. https://doi.org/10.1111/j.1083-6101.2011.01551.x

Zarouali, B., Strycharz, J., Helberger, N., & de Vreese, C. (2022). Exploring people's perceptions and support of data-driven technology in times of COVID-19: The role of trust, risk, and privacy concerns. *Behaviour & Information Technology, 41*(10), 2049–2060. https://doi.org/10.1080/0144929X.2021.2022208

About the authors

Peter Neijens is an Honorary Fellow and Emeritus Professor at ASCoR. He held the Chair in Media and Persuasion at the University of Amsterdam until 2019. His research focuses on persuasive communication, in particular media use and effects in advertising and public opinion. He was co-founder and scientific director of ASCoR from 1998 to 2005 and of the Graduate School of Communication from 2009 to 2013. Email: P.C.Neijens@uva.nl

Patti M. Valkenburg is a Distinguished University Professor at ASCoR. Her research focuses on the social and emotional effects of (social) media on youth and adults. She is particularly interested in theorising, studying, and demonstrating how individuals differ in their susceptibility to the effects of (social) media. E-mail: P.M.Valkenburg@uva.nl

Part II.

Empirical Findings and Theoretical Insights

3. Communicating Delegitimisation: Political Information and Challenges to Democracy

Michael Hameleers, Emily Gravesteijn, Linda Bos, and Alessandro Nai

Abstract

Political information that is disproportionately negative and uncivil may increase cleavages in society, fuel cynicism, and erode a shared understanding of factual information. ASCoR's focus on delegitimising communication has made an important contribution to our understanding of the nature, context, and potential democratic consequences of such potentially harmful information. Especially in digital settings, information that is at odds with democratic principles and objectivity may be disseminated at a high pace, whilst bypassing gatekeeping and fact-checking routines of established media. Although we conclude that we should be worried about delegitimising communication, discussions about pressing issues such as disinformation, polarisation, and echo chambers may also be harmful, as they can fuel overall and disproportionate cynicism toward (political) communication.

Keywords: polarisation, populism, disinformation, echo chambers, negative campaigning

Introduction

Many trends in political communication and journalism highlight negative forms of information that oppose democratic principles of truth-seeking, objectivity, and balance. These trends that can potentially undermine

Hameleers, M., Gravesteijn, E., Bos, L., & Nai, A. (2024). Communicating delegitimisation: Political information and challenges to democracy. In T. Araujo & P. Neijens (Eds.), *Communication research into the digital society: Fundamental insights from the Amsterdam School of Communication Research* (pp. 53-67). Amsterdam University Press. https://doi.org/10.5117/9789048560592_ch03

democratic values are accelerated by developments in digital media, which allow for the platforming and fast spread of information that has a delegitimising impact on society. In line with this trend, ASCoR researchers have mostly focused on analysing forms of communication that are potentially harmful for democracy. Among other things, we have focused on the centrality of conflict and negativity in political campaigns, the content and style of delegitimising populist communication, accusations of bias and other forms of criticism towards established media institutions, and the dissemination of mis- and disinformation. We consider these forms of information as potentially problematic as they may stand in the way of a rational exchange of ideas in society. Although negativity, conflict, and an appeal to ordinary people may be accepted or prevalent forms of political campaigning, we argue that the disproportionate use of such communication and its digital amplification may threaten democratic outcomes. In this chapter, we, therefore, use delegitimising communication as an overarching term to synthesise the research foci of ASCoR researchers studying political communication and journalism. This umbrella term refers to any form of communication that may threaten core deliberative and representative democratic values, such as a diversity of viewpoints, dialogue, and a constructive exchange of arguments between disagreeing groups in society, and which may be directed at democratic institutions that embody these values. It can also involve communication that is uncivil, one-sided, and (intentionally) false. We refer to it as delegitimising as it may attack or threaten conventional information and political institutions, such as mainstream and public media, and fuel cynical and distrusting views in democracy.

Because delegitimising communication may attack conventional truths and reinforce cleavages in society (e.g., van Dalen, 2021), we consider such information as problematic. Considering that the media are central players in conveying such information to citizens, a media focus on delegitimising information is crucial. Against this backdrop, it is important to map the role of (social) media in platforming and amplifying delegitimising communication, which has been a central theme in the empirical endeavours of ASCoR researchers. Here, we specifically pay attention to social and digital media as receptive platforms for delegitimising communication. As social media are not restricted by gatekeepers and the journalistic norms and practices of traditional media, they may play a central role in the communication of delegitimising communication. This has, for example, been established in research on populist communication (Hameleers & Schmuck, 2017) and negative campaigning (Petkevic & Nai, 2022). In the absence of traditional gatekeepers, various types of uncivil, hostile, negative, and delegitimising messages may find their way to the audience in an unfiltered manner.

Here, we should also consider the negative consequences of delegitimising narratives for journalism as a profession, as well as for the news institutions journalists are affiliated with. In line with the abundance of counterfactual narratives and disinformation online, journalists and other knowledge disseminators are often accused of spreading fake news, whilst they also face severe threats on their safety. Likewise, current debates about "alternative facts," fake news claims, and accusations of left-wing bias targeted at mainstream news outlets threaten to undermine the role of established media. Severe threats towards journalists are notable especially outside of the Western countries that we usually study (Balod & Hameleers, 2021). Thus, next to the threats of delegitimising communication for democracy, such information may also harm the discipline of journalism and the freedom with which journalists may cover democratic processes.

In the sections that follow, we will outline the approaches used by ASCoR researchers in mapping delegitimising communication. After this, we will present the main contributions stemming from these approaches. Here it should be noted that the findings and conclusions documented in this chapter mostly reflect general patterns across different settings in Western Europe and the United States. Whenever relevant, we will explicate to what extent findings are generalizable across Western countries, or whether country-specific differences are relevant to consider. We will end with an overview of directions for future research, and our ambitions to more comprehensively map delegitimising communication in a changing (digital) media landscape. Although we deal with a broad topic—delegitimising communication—we believe that many different research foci fall under this form of information. Specifically, in this chapter, we will review research that focuses on information that (1) emphasises negativity and conflict, (2) attacks the established order, and herewith (3) potentially threatens or undermines democracy.

Empirical findings

In this section, we discuss the different approaches taken by ASCoR researchers to map delegitimising communication. Overall, our research has mostly focused on a quantitative and/or automated assessment of delegitimising communication. Thus, most of the research findings presented here are not based on interpretative methods; rather, they paint a general picture of trends in delegitimising communication in big data sets.

A dominant line of research within delegitimising communication focuses on populist communication. That is, communication supporting the idea that society is "ultimately separated into two homogeneous and antagonistic groups, 'the pure people' versus 'the corrupt elite,' and which argues that politics should be an expression of the *volonté générale* (general will) of the people" (Mudde, 2004, p. 543). Although references to the people and their will are not delegitimising on their own, populism's emphasis on a moral divide between the people and "corrupt" elites is regarded as a delegitimising social identity frame (e.g., Bos et al., 2020). Especially the last element in populist ideology—the claim that politics should be an expression of the will of the people—is delegitimising in nature. It delegitimises all claims and issue stances that are, in the eye of the populist, not in line with "what the people want," legitimising an anti-pluralist or monist perspective on politics and democracy. Various ASCoR researchers studied the idiosyncrasy in communication by populist parties and politicians, showing that they are more populist in rhetoric and style (Bos & Brants, 2014) and more negative, incorporating more attacks and fear messages (Nai, 2021), including the most harmful types of conflict frames (van der Goot et al., 2022), and spreading more misinformation (Hameleers & Minihold, 2022). Yet, these studies mostly refer to research done in Western Europe, and herewith may not be directly transferable to other settings, such as bipartisan or more polarised settings. Additionally, many of these studies rely on research on populist radical right parties, and knowledge is much less saturated on the communication elements present in communication by left-wing populist parties.

Focusing on the dissemination of populist messages, Bos and Brants (2014) used a quantitative content analysis to map the degree of populist communication across different media over 20 years in the Netherlands—also distinguishing between populism used by the media and attention to populist ideas by journalists. Similar to Hameleers et al. (2019), they found that populist ideas expressed by traditional news media are extremely rare—indicating that news media may not directly contribute to the platforming of populist ideas that potentially undermine democracy. Yet, this potentially works differently in countries where more populist and hyper-partisan news outlets are part of the mainstream media landscape, such as the United States. Recently, in the Netherlands, we also see that hyper-partisan news outlets such as the populist radical right broadcaster Ongehoord Nederland give a podium to populist ideas on the public broadcaster.

Findings from studies in the United States on delegitimising forms of disinformation that attack established truths confirm some of the patterns found in Western European studies on populism. Traditional media most likely merely play a role as amplifiers of newsworthy deceptive information, for example, by repeating the false claims of political actors expressed online (Hameleers, 2021) and covering negative campaigns as well as candidates with a "darker" personality profile (e.g., Nai & Maier, in press). Traditional media may thus be part of the problem as accelerators of problematic forms of communication but often refrain from using such coverage themselves. Thus, looking at populist communication, disinformation, and negative campaigning, most research findings indicate that traditional (non-partisan and quality) media and journalists do not communicate in delegitimising manners themselves, although they may give a favourable and uncritical platform to political actors that are delegitimising. By uncritically offering a stage to such communication, disproportionally negative, uncivil, and untrue messages may be legitimised.

In this light, the increasing focus on digital media by ASCoR researchers may be explained as digital platforms offer an uncritical and unfiltered stage for the communication of delegitimising communication. On digital platforms, and in social media in particular, selective exposure to congenial or like-minded views (e.g., Trilling et al., 2017; Vermeer et al., 2020), the experience of like-mindedness due to "fringe" bubbles (e.g., Möller et al., 2018), and the AI-driven curation of content are specifically important to consider. They are assumed to platform and contribute to the fast-paced dissemination of delegitimising content. That is, targeted misinformation in the form of deepfakes are assumed to spread and flourish online (Dobber et al., 2021), and AI generates like-minded fringe bubbles of citizens distancing themselves from democratic and legitimate information.

Yet, despite the many worrisome popular discussions on these issues, ASCoR researchers have also relativised important concerns on the threats posed by delegitimising communication and its online dissemination. In that regard, Möller et al. (2018) emphasised that filter bubbles and echo chambers are not as pervasive as often suggested: The problem of such isolated spaces is rather small and extreme groups that may isolate themselves from the other side by only engaging with attitude-reinforcing communication are rare. Wojcieszak et al. (2023) show that due to the very limited fraction of news in people's online activities, the impact of online news exposure on polarisation is absent. On a more worrying note, a recent study by Simon et al. (2022) shows that extremist and delegitimising content can spread quite easily on a relatively open discussion platform such as Telegram. This reveals that when aiming to understand the content of delegitimising communication, the context matters: Such communication may not be found in the general media ecology but is more likely to be present in specific digital spaces, and may be disseminated and shared by people who are distrusting toward established institutions and conventional truths.

This does not mean that all concerns about delegitimising information are not backed up by empirical evidence. In online settings, populist communication is often communicated in hostile and uncivil ways (Nai, 2021)—even resonating with hate speech and the exclusion of out-groups (Hameleers, 2021). And such "darker" forms of elite rhetoric can, under certain circumstances, foster (support for) political violence (Nai & Young, 2023). Likewise, although disinformation and deepfakes are less present than assumed, they can be communicated and shared by distrusting segments of the audience (Dobber et al., 2021)—and their online targeting to vulnerable segments of society may indicate that they contribute to polarisation (Hameleers, 2021). And, most importantly, the fact that journalists rarely engage in negative, populist, and uncivil communication themselves does not rule out the indirect effects these campaign tactics have on journalism and trust in journalistic information. Dark communication that disproportionally focuses on conflict and negativity—if not too dark—is still considered newsworthy, and the populist representative claim is especially difficult to ignore. Here journalists find themselves in a Catch-22: neglecting populists serves their anti-elite and anti-media argument, while covering them—even in a critical way—legitimises their political existence (e.g., Bos et al., 2011; Vliegenthart et al., 2012).

Here, we should also underline the context of delegitimising communication: the prevalence of (societal and mediatised) discussions on issues such as fake news, filter bubbles, populism, conspiracies, and misleading information may reinforce distrust and confusion in society. For example, because fake news is used as a delegitimising label, affecting the perceived credibility of journalistic news stories (Bos et al., 2023), creating uncertainty in whom to trust (Brosius et al., 2022). In line with this, trends toward delegitimising communication may pose a severe threat on journalism as a profession—when facing accusations of fake news, bias, and a distortion of reality, journalists may no longer feel equipped to enact their crucial role perceptions as watchdogs beyond the dissemination of factual information (Balod & Hameleers, 2021). Future research should equally consider whether—and, if so, by whom and for which reasons—accusations of spreading misinformation or promoting "alternative facts" are used strategically, for example, to attack political opponents in a negative campaigning setting. To the best of our knowledge, no existing research has (yet) investigated such crucial intersection between promotion of disinformation and strategic accusations to do so for electoral purposes. Generally, we are thus in need of research that maps the motivations behind the communication of delegitimising information: What gains are sought, and what benefits are foreseen when political opponents are attacked or polarised cleavages emphasised?

In accordance with these developments, we see an increasing politicisation of public media, for example, through accusations of bias directed towards public news outlets such as public broadcasters in Europe. While public media hold a neutral function in democratic societies, claims that these institutions would supposedly not be neutral, but biased towards certain ideologies or political groups within society may seriously impair their legitimacy, and thus enhance an erosion of trust in media and democratic institutions more broadly (Brosius et al., 2022).

Contributions to theory and practice

By mapping the nature and context of delegitimising communication, ASCoR researchers have made several important contributions to theory and practice. The relativisation of the threats associated with echo chambers and filter bubbles have contributed to our understanding of the mechanisms of selective exposure and the role technological developments play in the curation of information flows (Möller et al., 2018, but also see Simon et al., 2022). Overall, our research has shown that delegitimising communication is often fragmented in nature, platformed on social media, and amplified by the affordances of the digital society. Yet, not everyone is exposed to such communication, and we cannot state that delegitimising communication has dominated political communication in the current information landscape.

Our extensive research agenda on populism, negative campaigning, and conflict framing has revealed various important theoretical insights on typologies of populism (Hameleers, 2021) and their delegitimising mechanisms (Bos et al., 2020; Hameleers et al., 2017). Related, we have contributed to a more refined understanding of conflict frames (Bartholomé et al., 2018; van der Goot et al., 2022)—and on how these different forms of "aggressive" elite rhetoric interplay, for instance, by focusing on the use of negative campaigning, populist appeals, and negative emotionality across the world, and their joint effects on election results and media attention (Gerstlé & Nai, 2019; Maier & Nai, 2020). These theoretical and empirical contributions spearheaded more comprehensive insights related to the conditions under which certain forms of delegitimising communication may undermine democratic principles, and herewith also require interventions. Hence, by indicating that some platforms, sources, and groups are more likely to disseminate delegitimising content, and by mapping for which sub-groups in society such messages may be problematic, we have contributed to a better understanding of the granular and fragmented nature of problematic forms of communication. More specifically, our research has contributed to a more refined understanding of the styles used to frame populist ideas, and the extent to which populist communication that stresses a divide between "us and them" may harm democratic principles such as civility and mutual respect. Research on the nature of disinformation has revealed how the truth—an important foundation of representative democracy—may be undermined and delegitimised deliberately. Studies on conflict framing and negative campaigning similarly contribute to our understanding of how the attack on out-groups and incivility may contribute to a context where the rational exchange of ideas between disagreeing citizens is undermined.

Research by ASCoR has also contributed conceptually to understanding various forms of delegitimising attributions towards the media. Qualitative, explorative work by ASCoR researchers on how bias accusations towards public media are constructed by political elites and opinion leaders has generated important insights into the extent to which academic debates about media bias correspond to the actual debates on media bias (Gravesteijn et al., 2023). This study exemplified the broader nature and directionality of bias accusations towards public service media, and distinguished it from other forms of media criticisms. It is important to understand what bias accusations citizens are exposed to, as it potentially affects their trust in media and general opinion of news outlets (Bos et al., 2023). Moreover, a disparity between understanding bias in academic debates and public debates will potentially impair addressing and remedying perceptions of biased news outlets amongst citizens, and thus will not effectively contribute to bolstering perceived legitimacy of public media in democratic societies.

On a more general level, ASCoR researchers have advanced theories on how the setting of digital affordances (i.e., social media, algorithms, AI) have changed the nature of delegitimising communication: It has become more personalised, tailored, uncivil, whilst empowering different voices to express themselves via online platforms (Simon et al., 2022). Although the main focus of our work has been on negative ramifications for democracy, social media and technological developments should not be regarded as bad in their own right. While ASCoR researchers focus on positive implications of digital media less, the opportunity for community formation, information sharing, and cross-cutting information exposure can be contrasted to the problematic context of delegitimising communication. However, by relativising the omnipresent concerns on the allegedly destabilising impact of echo chambers, filter bubbles, deepfakes, misinformation and populism by the media, we have contributed to a more nuanced understanding of the democratic consequences of delegitimising communication in a digital information ecology. Although we highlight the problematic potential of delegitimising communication, we also emphasise that such communication is specifically harmful for specific groups of society, such as citizens with cynical views on politics or people who are already inclined to select delegitimising information (Bos et al., 2020; Hameleers et al., 2018).

On a more general level, the insights presented in this chapter lay the foundation of a theory that integrates various forms of problematic and negative forms of conflictive communication under the framework of delegitimising communication. Hence, moving beyond isolated concepts such as disinformation, populism, or negative campaigning, delegitimising communication revolves around negatively biased information that may—both intentionally and unintentionally—harm democracy by attacking conventional truths, democratic institutions or the established political order. Such communication is at odds with truth telling, balance, and objectivity. Its causes may be psychological (i.e., people's preference for like-minded views), technological (i.e., the platforming of delegitimising communication online), and sociopolitical (i.e., the success of populist movements that have mainstreamed delegitimising communication). A comprehensive model of the context, content, causes and potential democratic consequences of delegitimising communication makes an important theoretical contribution to communication science. Such a more comprehensive framework may help us to understand under which conditions delegitimising communication may be expressed, and what its intended outcomes are. At this stage, ASCoR researchers have mostly focused on the content and effects of this theoretical framework. Future endeavours may focus more on the causes and contexts, for example, through interpretative methods that reveal the motivations for communicators to communicate in delegitimising ways, or to disseminate such information through social media platforms.

To extend the impact of our research agenda on delegitimising communication, we also translate our findings to practical implications and recommendations. Our research on mis- and disinformation has, for example, important implications for the design of effective interventions. Among other things, our findings indicate that fact-checks are effective, but only when they are selected by audience segments that were actually persuaded by false information (Hameleers & van der Meer, 2020). Hence, as most people select fact-checks that confirm their opposition to disinformation, it is crucial to assess how corrective information can reach people that tend to believe the refuted false claims in the first place. In addition, our research indicates that media literacy interventions only work effectively if they can enhance trust in factually accurate information (van der Meer et al., 2023). These empirical conclusions are widely shared with practitioners in journalism and media literacy education. To offer a few examples, ASCoR researchers are part of the EDMO network that connects academic insights on fact-checking to actual suggestions for interventions implemented by fact-checkers in the Netherlands and Belgium.[1] We are also developing new media literacy tools that can be used as a browser plug-in to warn people about dubious information. On a more general level, we aim to connect our research to various stakeholders in journalism and political communication to inform the development of potential interventions that could alleviate the threats of delegitimising information. At the same time, however, we also offer evidence for the omnipresent, and at times inflated, concerns about dark and undemocratic forms of communication that allegedly dominate online.

Next steps

As explicated in the previous sections, ASCoR focuses predominately on the negative aspects of information flows; as such, the "dark" side of communication tends to prevail in our ongoing lines of research. However, the affordances of online media may also offer opportunities for civil, democratic, and diverse exchanges between citizens, as people with different views may come together at a marketplace of ideas. Hence, online, people may be confronted more with different views as compared to their direct offline environment that tends to be more homogenous (Trilling et al., 2017). In addition, the developments of AI, algorithms and engaging personalised communication can also be used as solutions to the problems facing journalism and political communication. Our future research agenda could make solutions-oriented journalism and political communication more central, herewith offering evidence-based suggestions for how the potential problems of delegitimising communication and growing levels of distrust in political institutions and information may be remedied. To offer a few examples, research on disinformation could focus more on the feasibility of media literacy programmes, and research on polarisation could investigate the potential of depolarising forms of online communication.

1 https://benedmo.eu/.

Another avenue for future research concerns a better understanding of the causes and context of delegitimising communication. The focus on quantitative and computational methods has merits, but it may overlook the perceptions of people producing and consuming delegitimising information. In addition, we know markedly little about why certain delegitimising communication is expressed, which makes research on potential causes crucial. For example, we know little about the conditions under which false information is disseminated unintentionally (i.e., due to honest mistakes) or intentionally (i.e., to steer political perceptions). To advance this line of research, we could focus more on qualitative and interpretative methods, such as in-depth interviews with citizens exposing themselves to delegitimising communication, politicians spreading disinformation or negative information, or journalists who are affected by delegitimising attacks. Generally, the changing role of journalists in a time of constant delegitimising attacks, and the potential undermining role of alternative forms of media online, deserves more attention in our future research. Hence, especially as we theorise that delegitimising information is often intentional and goal-directed, focusing on just mapping content features may overlook underlying motivations to spread or create delegitimising narratives. Our ambition is to rely on more systematic interpretative qualitative work to unravel the drivers of delegitimising communication. In addition, we also consider it important to further extend our research agenda to studying the (potentially harmful) effects of delegitimising narratives and labels on citizens—specifically in terms of how they view journalists and news outlets to which these delegitimising labels are directed, and whether exposure to these delegitimising narratives affect these perceptions.

Finally, we would like to highlight the importance of comparative research, or research conducted in a diversity of contexts. The conclusions we reach about the threats of delegitimising communication mainly stem from empirical evidence collected in Western Europe or the United States. This means that we cannot automatically make inferences about the nature and implications of delegitimising communication across the globe. Arguably, journalists in less free countries with high levels of state ownership in the media landscape face different challenges than journalists who can work independently. Likewise, the nature of disinformation narratives is very different when a state is pushing propaganda narratives to silence the opposition compared to a country where disinformation mainly comes from foreign states, such as Russian (via the Internet Research Agency). More comparative research designs, or the extension of our geographical lenses in empirical research, may offer more nuanced insights of the role of contexts, such as media and political systems, in the expression, platforming, and perceptions of delegitimising communication.

Conclusions

In this chapter, we have defined delegitimising communication as negative information that focuses on conflict and societal oppositions. Here, reflecting the work of ASCoR researchers, we especially focused on populism, negative campaigning, and disinformation. Such communication may intentionally and unintentionally harm democracy by fuelling polarisation, cynicism, and distrust in established knowledge. In recent years, delegitimising communication has obtained an important epistemic dimension: Negative messages may not only stress divides between groups based on ideological or partisan cleavages, but also relate to cleavages between opposed truths and the delegitimisation of established information. As a major theoretical contribution, we believe that a focus on delegitimising communication may bring together various forms of harmful content, and this approach allows for an assessment of the context, causes, and effects that such communication may have across democracies. Hence, rather than studying negative forms of communication in isolation, a comprehensive theoretical account of delegitimising communication may point to shared root causes, mechanisms behind effects, and solutions that help to make societies more resilient to delegitimising content.

References

Balod, H. S. S., & Hameleers, M. (2021). Fighting for truth? The role perceptions of Filipino journalists in an era of mis- and disinformation. *Journalism*, *22*(9), 2368–2385. https://doi.org/10.1177/1464884919865109

Bartholomé, G., Lecheler, S., & de Vreese, C. (2018). Towards a typology of conflict frames: Substantiveness and interventionism in political conflict news. *Journalism Studies*, *19*(12), 1689–1711. https://doi.org/10.1080/1461670X.2017.1299033

Bos, L., & Brants, K. (2014). Populist rhetoric in politics and media: A longitudinal study of the Netherlands. *European Journal of Communication*, *29*(6), 703–719. https://doi.org/10.1177/0267323114545709

Bos, L., Egelhofer, J. L., & Lecheler, S. K. (2023). Short but critical? How "fake news" and "anti-elitist" media attacks undermine perceived message credibility on social media. *Communication Research*. Advance online publication. https://doi.org/10.1177/00936502231178432

Bos, L., Schemer, C., Corbu, N., Hameleers, M., Andreadis, I., Schulz, A., Schmuck, D., Reinemann, C., & Fawzi, N. (2020). The effects of populism as a social identity frame on persuasion and mobilisation: Evidence from a 15-country experiment. *European Journal of Political Research*, *59*(1), 3–24. https://doi.org/10.1111/1475-6765.12334

Bos, L., van der Brug, W., & de Vreese, C. (2011). How the media shape perceptions of right-wing populist leaders. *Political Communication*, *28*(2), 182–206. https://doi.org/10.1080/10584609.2011.564605

Brosius, A., Hameleers, M., & van der Meer, T. G. (2022). Can we trust measures of trust? A comparison of results from open and closed questions. *Quality & Quantity*, *56*(5), 2907–2924. https://doi.org/10.1007/s11135-021-01250-3

Dobber, T., Metoui, N., Trilling, D., Helberger, N., & de Vreese, C. (2021). Do (micro-targeted) deepfakes have real effects on political attitudes? *The International Journal of Press/Politics*, *26*(1), 69–91. https://doi.org/10.1177/1940161220944364

Gerstlé, J., & Nai, A. (2019). Negativity, emotionality and populist rhetoric in election campaigns worldwide, and their effects on media attention and electoral success. *European Journal of Communication*, *34*(4), 410–444. https://doi.org/10.1177/0267323119861875

Gravesteijn, E., van Elsas, E., & Gattermann, K. (2023, January 10). *What bias? Understanding bias accusations towards Public Service Media in the Netherlands.* OSF. https://doi.org/10.17605/OSF.IO/8YCH2

Hameleers, M. (2021). *Populist disinformation in fragmented information settings: Understanding the nature and persuasiveness of populist and post-factual communication.* Routledge. https://doi.org/10.4324/9781003194668

Hameleers, M., Bos, L., & de Vreese, C. H. (2017). "They did it": The effects of emotionalized blame attribution in populist communication. *Communication Research*, *44*(6), 870–900. https://doi.org/10.1177/0093650216644026

Hameleers, M., Bos, L., & de Vreese, C. H. (2018). Selective exposure to populist communication: How attitudinal congruence drives the effects of populist attributions of blame. *Journal of Communication*, *68*(1), 51–74. https://doi.org/10.1093/joc/jqx001

Hameleers, M., Bos, L., & de Vreese, C. H. (2019). Shoot the messenger? The media's role in framing populist attributions of blame. *Journalism*, *20*(9), 1145–1164. https://doi.org/10.1177/1464884917698170

Hameleers, M., & Minihold, S. (2022). Constructing discourses on (un)truthfulness: Attributions of reality, misinformation, and disinformation by politicians in a comparative social media setting. *Communication Research*, *49*(8), 1176–1199. https://doi.org/10.1177/0093650220982762

Hameleers, M., & Schmuck, D. (2017). It's us against them: A comparative experiment on the effects of populist messages communicated via social media.

Information, Communication & Society, 20(9), 1425-1444. https://doi.org/10.1080/1369118X.2017.1328523

Hameleers, M., & van der Meer, T. G. A. (2020). Misinformation and polarization in a high-choice media environment: How effective are political fact-checkers? *Communication Research*, *47*(2), 227–250. https://doi.org/10.1177/0093650218819671

Maier, J., & Nai, A. (2020). Roaring candidates in the spotlight: Campaign negativity, emotions, and media coverage in 107 national elections. *The International Journal of Press/Politics*, *25*(4), 576–606. https://doi.org/10.1177/1940161220919093

Möller, J., Trilling, D., Helberger, N., & van Es, B. (2018). Do not blame it on the algorithm: An empirical assessment of multiple recommender systems and their impact on content diversity. *Information, Communication & Society*, *21*(7), 959–977. https://doi.org/10.1080/1369118X.2018.1444076

Mudde, C. (2004). The populist zeitgeist. *Government and Opposition*, *39*(4), 541–563. https://doi.org/10.1111/j.1477-7053.2004.00135.x

Nai, A. (2021). Fear and loathing in populist campaigns? Comparing the communication style of populists and non-populists in elections worldwide. *Journal of Political Marketing*, *20*(2), 219–250. https://doi.org/10.1080/15377857.2018.1491439

Nai, A., & Maier, J. (In press). *Dark politics: The personality of politicians and the future of democracy.* Oxford University Press.

Nai, A., & Young, E. L. (2023). *Blood on the campaign trail: Negative campaign rhetoric and political violence* [Manuscript submitted for publication]. Amsterdam School of Communication Research, University of Amsterdam.

Petkevic, V., & Nai, A. (2022). Political attacks in 280 characters or less: A new tool for the automated classification of campaign negativity on social media. *American Politics Research*, *50*(3), 279–302. https://doi.org/10.1177/1532673X211055676

Simon, M., Welbers, K., Kroon, A. C., & Trilling, D. (2022). Linked in the dark: A network approach to understanding information flows within the Dutch Telegramsphere. *Information, Communication & Society.* Advance online publication. https://doi.org/10.1080/1369118X.2022.2133549

Trilling, D., van Klingeren, M., & Tsfati, Y. (2017). Selective exposure, political polarization, and possible mediators: Evidence from the Netherlands. *International Journal of Public Opinion Research*, *29*(2), 189–213. https://doi.org/10.1093/ijpor/edw003

van Dalen, A. (2021). Rethinking journalist–politician relations in the age of populism: How outsider politicians delegitimize mainstream journalists. *Journalism*, *22*(11), 2711–2728. https://doi.org/10.1177/1464884919887822

van der Goot, E., Kruikemeier, S., de Ridder, J., & Vliegenthart, R. (2022). Online and offline battles: Usage of different political conflict frames. *The International Journal of Press/Politics*. Advance online publication. https://doi.org/10.1177/19401612221096633

van der Meer, T. G., Hameleers, M., & Ohme, J. (2023). Can fighting misinformation have a negative spillover effect? How warnings for the threat of misinformation can decrease general news credibility. *Journalism Studies*, *24*(6), 803–823. https://doi.org/10.1080/1461670X.2023.2187652

Vermeer, S., Trilling, D., Kruikemeier, S., & de Vreese, C. (2020). Online news user journeys: The role of social media, news websites, and topics. *Digital Journalism*, *8*(9), 1114–1141. https://doi.org/10.1080/21670811.2020.1767509

Vliegenthart, R., Boomgaarden, H. G., & van Spanje, J. (2012). Anti-immigrant party support and media visibility: A cross-party, over-time perspective. *Journal of Elections, Public Opinion & Parties*, *22*(3), 315–358. https://doi.org/10.1080/17457289.2012.693933

Wojcieszak, M., de Leeuw, S., Menchen-Trevino, E., Lee, S., Huang-Isherwood, K. M., & Weeks, B. (2023). No polarization from partisan news: Over-time evidence from trace data. *The International Journal of Press/Politics*, *28*(3), 601–626. https://doi.org/10.1177/19401612211047194

About the authors

Michael Hameleers is an Assistant Professor of Political Communication & Journalism at ASCoR. His research interests include (right-wing) populism, disinformation, and selective exposure. He has published extensively on (visual) disinformation and the effects of populism. E-mail: M.Hameleers@uva.nl

Emily Gravesteijn is a PhD student in Political Science and Political Communication at Radboud University and ASCoR. Her research focuses on understanding why citizens perceive a bias in news from public service media. E-mail: Emily.Gravesteijn@ru.nl

Linda Bos is an Associate Professor and Programme Group Director of Political Communication & Journalism at ASCoR. She is an expert in the field of populist political communication. Her broader line of research centres on the relation between political elites, media, and voters. E-mail: L.Bos@uva.nl

Alessandro Nai is an Associate Professor of Political Communication & Journalism at ASCoR. His research focuses on the dark sides of politics, the use of negativity and incivility in election campaigns in a comparative perspective, and the (dark) personality of political figures. E-mail: A.Nai@uva.nl

4. Disenchantment with Political Information: Attitudes, Processes, and Effects

Alessandro Nai, Susan Vermeer, Linda Bos, and Michael Hameleers

Abstract

This chapter investigates nefarious dynamics of the "demand" side of political communication, that is, the negative side of information processing at the individual level. It focuses on sets of mechanisms at the two "ends" of individual information processing: First, looking first at "downstream" effects, the chapter discusses recent and current research by ASCoR about the potentially nefarious effects of exposure to dark and delegitimising political information, e.g., in terms of depressed voter turnout and increased polarisation. Second, looking at "upstream" mechanisms, the chapter discusses evidence that important segments of the population have pre-existing negative attitudes towards political information itself. A final section opens up on the challenges, both theoretical and methodological, that our discipline is facing.

Keywords: persuasion, negativity, cynicism, populism, news avoidance

Introduction

In an ideal world, political information—that is, any type of information from or about the political world (from news media coverage of policy implementation to campaign materials published by parties, and everything in between)—would only convey relevant and meaningful material, be universally accessible and easy to process, retrieve and sample, have only

Nai, A., Vermeer, S., Bos, L., & Hameleers, M. (2024). Disenchantment with political information: attitudes, processes, and effects. In T. Araujo & P. Neijens (Eds.), *Communication research into the digital society: Fundamental insights from the Amsterdam School of Communication Research* (pp. 69-86). Amsterdam University Press. https://doi.org/10.5117/9789048560592_ch04

socially desirable effects, and is equally appreciated by all citizens (Nai, 2019). Such patterns of communication would be in line with the ideals of a deliberative democracy, where debates are based on an elaboration of substantive arguments that are rooted in factually accurate and balanced information diets.

In reality, we have moved far away from this (certainly naïve) ideal. Political information can convey nefarious meaning, be processed in a biased and skewed way, be promoted following perverse incentives, enhance misperceptions, and generally be societally harmful. With this in mind, this chapter focuses on these "dark" aspects of political information exposure and processing, and it showcases the extensive research that ASCoR scholars have been engaged in. Providing a comprehensive overview of all these dysfunctional and negative dynamics involved in (political) information processing is a daunting task, but luckily not one that this chapter tackles. Rather, we decided to focus here on two substantive lines of research, investigating the two main sides of the political information process: its effects and exposure to it. First, we discuss how recent research by ASCoR scholars have tackled the fundamental question of the "effects" of information—both news content and direct communication from political actors—focusing on the nefarious consequences of "dark" political information for individual attitudes and behaviours. Can political information really persuade us? Disenfranchise us and push us towards apathy and cynicism? Or even activate the aggressive and violent side in us? Second, we turn to the negative attitudes towards political information. Are there segments of the public that particularly dislike or distrust the news? Is this perhaps associated with the broader dynamics of populism? In a context in which interest and trust in news is declining around the world (Newman et al., 2022), understanding the individual roots of negative perceptions of political information becomes increasingly central.

Empirical findings: Dark information, its effects, and information avoidance

Conceptually, our overarching claim is that current dysfunctions within the broad process of reception and processing of political information could stem from two interrelated sets of mechanisms: (1) the fact that political information is often framed negatively (e.g., focusing on the darker and more uncivil side of politics), which likely has detrimental effects for subsequent attitudes and behaviours among those exposed to it ("downstream" effects), and (2) the fact that often a negative stance towards political information itself exists from those who might be exposed to it ("upstream" effects). This section discusses recent and current research by ASCoR members directly tackling these two main sets of mechanisms.

When it comes, first, to the effects of political communication, early studies were relatively sceptical about the capacity of (political) information to alter people's mind. Research paradigms developed in the wake of the Second World War suggested the presence of rather "minimal" effects of exposure to information for the society at large, stressing in particular the fundamental role of deep predisposition, values, and socialisation to explain voting patterns. Without discounting the epistemological and theoretical importance of these early models, what we consider a more fruitful approach is a detailed focus on individual consequences of exposure to political information. And, adopting such a narrower (but likely more accurate) focus, few would argue today that political information does not matter—for good and for bad. Indeed, recent research shows rather clearly that simple exposure to political messages can persuade—that is, move opinions and attitudes in line with the content of the information individuals were exposed to. Implementing a rather simple research design, where researchers exposed participants to tailored incongruent information on specific topics—that is, new information that was at odds with participants' initial stance of such topics—Nai et al. (2017, 2023) were able to show that between a quarter and 40% of respondents have unstable opinions: regardless of whether they initially have a positive or negative opinion about a given topic (e.g., implementing economic restrictions to curb climate warming), presenting them with arguments supporting the opposite stance makes them re-evaluate their beliefs and change their initial stance. To be sure, the magnitude of such changes was at times only marginal (e.g., moving from "strongly against" to "rather against" a given policy proposition), and people have at their disposal quite a large palette of strategies to resist those persuasive attempts (Valli & Nai, 2023b; van Reijmersdal et al., 2016). Yet, the fact that such changes exist in a large share of respondents and only after exposing them to a rather simple manipulation, strongly suggests that information shapes opinions quite directly—confirming well-known trends in political, social, and health communication (e.g., Neijens & de Vreese, 2009).

Information, then, can change people's minds and alter their attitudes and behaviours. With this in mind, a normatively important question that arises is whether such effects could be detrimental. Research on the nefarious consequences of exposure to information has, notably, amply discussed the existence of a "media malaise," according to which political attitudes such as efficacy, trust, and positive perceptions of the political system are depressed by exposure to the news media's portrayal of politics as a game or contest (e.g., Schuck, 2017)—and, as outlined above, this relationship between exposure and attitudinal consequences is likely self-reinforcing. Perhaps less consensual are the effects of non-mediated information, such as campaign materials sponsored by political actors for electoral purposes. Beyond its intended electoral effects, this type of information—omnipresent in competitive democracies—can have two types of nefarious consequences: depressing and polarising effects.

On the one hand, exposure to political messages from elites can depress the electorate and push them away from the voting booths. This is certainly the case of negative and uncivil campaign messages—that is, political advertising including attacks towards opponents ranging from policy criticism to mudslinging, character assassinations, and even outright ad hominem insults. Research on the dynamics of dark communication (i.e., negative campaigning and incivility) comes historically from the USA but has increasingly embraced a more international and comparative outlook (e.g., Maier et al., 2022; Nai, 2020; Walter & Vliegenthart, 2010). Importantly, some indication exists that exposing voters to negative and uncivil campaigns can depress their political engagement and creates a "gloomier" public, disinterested and disappointed with the political process. Exposed by a constant barrage of attacks and incivility, voters can lose sight of the ultimate importance of their political participation, and demobilise. Otto et al. (2020) carried out an experiment in three countries—the Netherlands, the UK, and Spain—testing whether civil and uncivil forms of mediated political conflict affected cynicism, political participation intentions, and support for the discussed policies. They found that while the first was not impacted, the latter two were depressed in all three countries after exposure to uncivil political conflict.

On the other hand, political messages—in particular, dark and delegitimising ones—can rile up passions and polarise the electorate—and even mobilise them to vote (Schuck et al., 2016). Exposure to populist messages—usually particularly hostile and uncivil (Hameleers, 2019; Nai, 2021)—tends to generate stronger emotional reactions than exposure to mainstream communication, and such emotions mediate or moderate the persuasiveness of populist appeals. In their experiment Hameleers, Bos, and de Vreese (2017) exposed respondents to messages framed as blame attributions towards the government or the EU (vs no blame attributed) while also manipulating anger and fear. The results of their study with 721

Dutch respondents shows that respondents in the fear conditions were more likely to accept populist blame attributions when the government was blamed. Specifically populist messages—signalling the distinction between the "good" people and the "bad" elite and other out-groups—can persuade voters to agree with a specific standpoint and mobilise them, when compared to messages that do not focus on any out-group (Bos et al., 2020). A 16-country experiment ($N = 7{,}286$) comparing (1) a message that was positive about the in-group of the people to (2) a message pitting the good people against the bad elite or (3) pitting them against immigrants or (4) both showed that the second anti-elitist message persuades voters to agree with the standpoint conveyed in the message. In addition, respondents scoring high on relative deprivation—believing that the resources available in society are not distributed in a fair way, and most importantly not in their interest—are more likely to be mobilised after exposure to this message. In line with this, many ASCoR researchers have studied the relationship between anti-immigrant news and anti-immigrant attitudes as well as populist radical right party support, showing a consistent relationship between the two (e.g., Boomgaarden & Vliegenthart, 2007, 2009; Damstra et al., 2021).

Most often information-induced anger and rage are directed towards political elites and other systemic "culprits" (Hameleers et al., 2017), but they can also have a "horizontal" target—the political out-group. In that sense populist messages do not necessarily polarise the electorate, underlining differences between voters, but most notably polarise the in-group of the people vis-à-vis the out-group of the elite, hereby creating collateral damage as a subsection of the electorate is often also perceived as part of the elite or establishment, left-wing, highly educated or "woke." The populist message in a sense legitimises negative attitudes towards the political out-group: it assumes that politics should be an expression of the general will of the "true" people. In line with this, a survey among a representative sample of Dutch voters ($N = 1{,}999$) shows that voters with stronger populist attitudes are less supportive of democratic norms, more intolerant of opposing views online, and more intolerant of political opponents (Bos et al., 2021). And this can lead to violence. A recent study showcases a disturbing positive association between negative and uncivil elite rhetoric and real-world instances of violent political incidents around election day (Nai & Young, 2023). In that sense, negative, uncivil, and/or populist rhetoric is not without consequence.

Above and beyond these nefarious "downstream" effects—that is, as a consequence of exposure to political information—current dysfunctions with the general process of information processing also come from "upstream" dynamics—that is, preceding exposure itself. Some people do not like political information or, quite simply, prefer to have little to do with it. This disenchantment with political information takes, broadly speaking, two interrelated forms: a generalised distaste (or indifference) towards news, or an ideologically driven normative distrust towards (some segments of) the information environment. Both forms of disenchantment broadly reflect an explicit or unconscious desire to avoid political information; where they differ is in the roots and mechanisms leading to such proclivity for avoidance. We discuss these two main forms of disenchantment below, before turning to research investigating the effects of exposure to (dark) information.

Some evidence exists of a generalised indifference and distaste towards political information. Some segments of the public simply "opt out" from specific types of news consumption. Based on a latent profile analysis with national survey data from the Netherlands (N = 2,833), Bos, Kruikemeier, and de Vreese (2016) find that 65.94% of the sample can be considered having a minimalist news diet, using the least amount of media, watching the fewest current affairs programmes, and reading virtually no newspapers. Additionally, citizens may decide to actively and consciously avoid the news (intentional news avoidance), because news use could have a negative effect on their well-being or because of information overload. By conducting two panel surveys in 2020 among Dutch citizens, de Bruin et al. (2021) show that this effect is even stronger during times of crisis (e.g., the COVID-19 pandemic). Conversely, some citizens may show low levels of news use without actively avoiding it (unintentional news avoidance), for example, because news consumption is not part of their daily routine. This generalised form of news avoidance exists along more "targeted" forms of avoidance, which tend to be ideologically motivated. Some voters explicitly reject (segments of) the informational environment due to ideological or normative considerations. Here, citizens' political attitudes and ideologies play a crucial role in their need for political information. The results of a panel study among Austrian citizens in 2020 indicate that citizens who have little trust in political institutions and political elites have less reason to stay informed about their actions through political news consumption (see, e.g., Schäfer et al., 2022).

In the context of news avoidance, trust in media also plays an important role. A lack of trust can stem from the perception that news does not provide neutral information about recent events, but rather that news organisations paint a biased picture and pursue their own agenda, or even promote and push disinformation (Hameleers, 2021). As a result, low media trust increases tendencies to turn away from news and turn towards social and non-mainstream news. Interestingly, Vermeer et al. (2022) show that, by using survey data from two waves from the Netherlands (with a baseline just before the outbreak of the COVID-19 pandemic), these behavioural patterns are related to trust in these alternative news media: the more citizens trust non-mainstream news and social media, the more they expose themselves to these alternative media and vice versa. This type of alternative news use can be consequential for political knowledge. A recent survey (Damstra et al., 2023) among 2,160 Swedish citizens shows that intentional news avoiders not only are less knowledgeable about politics in general, but also have the least accurate beliefs about contested issue domains. That is, they hold the most biased and misinformed beliefs about issues like climate change, vaccination, genetically modified organisms, crime, and immigration.

In line with this, the above mentioned study by Bos, Kruikemeier, and de Vreese (2016) shows that specifically right-wing populist voters have a more popular news diet consisting of the avoidance of established quality media and hard news formats (i.e., NOS) and the approach of popular information such as talk shows and the tabloid newspaper *De Telegraaf*. In addition, voters scoring high on a scale measuring relative deprivation (feeling unfairly treated by society) are more likely to select populist messages and subsequently report stronger populist attitudes (Hameleers et al., 2018). Likewise, citizens selecting into soft news diets as opposed to hard news tend to be more cynical of politics (Boukes & Boomgaarden, 2015). With populist messages being more readily available within these popular news diets, a populist news spiral is not unimaginable, with populist voters tuning more and more out of mainstream news and into more popular or hyper-partisan news. Yet, in their analysis of Dutch news diets, Bos et al. (2016) also find that the public broadcasting news is present in all media diets, something also found by Stier et al. (2020) in a more recent study analysing online information exposure. How these legacy news stories are subsequently processed is, however, most likely a different story. Thus, even though fringe or filter bubbles may not exist in the sense that disenchanted people still expose themselves to established news, it remains to be studied what motives people have when approaching mainstream sources.

Importantly, the relevance of studying these news consumption patterns should also be considered in light of the work showcasing its relevance for knowledge and political participation. In one of the first ASCoR linkage analyses, de Vreese and Boomgaarden (2006) investigated the relationship between media and political knowledge by combining panel surveys and news media content analyses in Denmark and the Netherlands, linking media content measures to individual attitudes through media exposure measures. The results show that exposure to news outlets with high levels of news content (e.g., public news broadcasts) contributes the most to knowledge gains and the intention to vote. Importantly, exposure to outlets with less political content has either no or very little positive effects. Based on these findings, they conclude that the effects of news media on knowledge and participation are more virtuous than vicious. Along the same lines, Elenbaas et al. (2014) find that a wider availability of political information in the media environment has a strong and positive impact on individual acquisition of political information—and that individuals more interested benefit more, but this difference disappears when information saturation is higher. Finally, Möller, Kühne, and de Vreese (2018) surveyed young voters (N = 994) around the 2014 European Parliament election, connected the survey data to a content analysis of the campaign (N = 769) and found that exposure to news in offline media had no significant impact on participation, whereas exposure to news in online media did positively affect turnout, suggesting that patterns of participation as a function of political news consumption strongly depend on the social and technological context in which information is accessed and shared.

Contributions to theory and practice

In the past decades the literature has expanded considerably towards a deeper and more nuanced understanding of the dynamics of individual disenchantment with political information—often spearheaded by cutting-edge contributions from ASCoR researchers. Taking stock of these advances and looking ahead at what's to come in this ever-expanding field, three outstanding issues are on the table.

First, from a theoretical standpoint, information processing and individual differences necessarily intersect. Taken separately, the evidence discussed above seems to suggest that the negative attitudes towards and negative consequences of political information exist as independent sets of mechanisms. This is, of course, an oversimplification and, indeed, strong evidence exists that individual factors and the content of political information jointly explain the (nefarious) effects of the latter. More specifically, a consensus seems to emerge that a non-negligible segment of the electorate seems to "like" dark and delegitimising rhetoric—or, at the very least, does not seem to be turned off by it. Research highlighting this trend focuses on a disparate sets of factors; for instance, voters high in tolerance for disagreement seem to be immune from the demobilising effects of incivility (Otto et al., 2020), while voters low in agreeableness are more attracted to the antiestablishment message of populist politicians (Bakker et al., 2021). Character attacks and incivility are more persuasive among respondents, respectively, high in psychopathy and low in agreeableness (Nai & Maier, 2021), and political news showcasing a conflict frame is more successful in mobilising voters low in conflict avoidance (Bjarnøe et al., 2020). While disparate, these factors seem to point towards the idea that it is more "aggressive" individuals—less likely to avoid conflict, with a "darker" personality profile, or quite simply scoring high in attitudinal predispositions towards behavioural aggressiveness—who seem to have a soft spot for aggressive political communication. Yet, such evidence remains at a preliminary stage, and a full-fledged framework about the moderating role of individual predispositions towards "dark" and delegitimising information remains an outstanding challenge.

Second, epistemologically, digital technologies, such as (new) social media platforms, have dramatically changed the way citizens are informed about political and societal issues. While social media platforms provide citizens with new opportunities to access information, express opinions, and participate in democratic processes, they can also undermine democracy by allowing polarising messages, conspiracy theories, extremist views, and unreliable information to spread easily, likely even more than less polarising, more nuanced perspectives. Here a recent study by Simon et al. (2022) using web mining, neural topic modelling, and social network analysis techniques on a dataset of 174 Dutch-language public Telegram chats/channels shows that over time, conspiracy-themed, far-right activist, and COVID-19–sceptical communities dominated the Dutch Telegramsphere of current affairs. These situations may push social media platforms to rely on automated tools to identify such content, which can result in algorithmic bias. Digital technologies also have an impact on trust perceptions, risk beliefs, and privacy concerns—the most pivotal issues in the digital age of information and communication—among citizens (Kruikemeier et al., 2020; Zarouali et al., 2022). As this may affect citizens' online behaviour and capacity to form and express political opinions, such findings contribute to an important and timely debate regarding the democratic legitimation of digital technologies. Digital technologies and the affordances of artificial intelligence have offered the opportunity for malicious actors to deceive the public in increasingly more realistic ways, for example, through the use of visual disinformation and deepfakes (Dobber et al., 2021). Hence, when citizens are deliberately misinformed, or confronted with multiple alternative truth claims on the same issue, their ability to make well-informed decisions based on a common truth is hampered. The role of digital technologies in exacerbating the processes described in this chapter is high on the agenda.

Third, we should consider the context of misinformation, disinformation, and polarisation in more depth (e.g., van der Meer et al., 2023). Although focusing on the effects of exposure to delegitimising false information is important, the threats of mis- and disinformation are also perceptual and discursive. Specifically, the increasing public attention to problems of disinformation has paved the way for the use of "fake news" and other delegitimising labels as political tactics, which may raise doubt in factually accurate information, confuse the electorate, or eventually contribute to the avoidance of all established information (Hameleers, 2021). The perception that society is polarised or that information is fake may have real consequences for trust in accurate information and democracy in general, and we should not overlook these side effects of emphasising threats in the information ecology. A recent study by Bos, Egelhofer, and Lecheler (2023) uses a factorial survey ($N = 715$) to study the impact of fake news and anti-elite cues in undermining news credibility. While the results lend support for a more optimistic view—showcasing that source and confirmation heuristics constitute the strongest influences on citizens' perceptions of the credibility of an online news message—the findings also show that the fake news cue and the anti-elite cue can decrease credibility if a news message is incongruent with voters' issue positions or the anti-elite cue is in line with their populist attitudes.

Next steps

The wealth of evidence unearthed by ASCoR research has helped unpack some of the most central mechanisms associated with dysfunctional processing of political information. Yet, as it is often the case, for each conceptual and empirical advance a new series of challenges stems. We highlight some of these challenges here, not with the aim of being comprehensive but rather to hint at some notable future directions towards which we will steer our collective endeavour.

Mounting evidence about the interrelations between individual predispositions and consequences of information processing raises a challenge of a more methodological nature: the existence of self-reinforcing spirals. While the existence of reinforcing spirals between exposure to information and political attitudes has been acknowledged in the past (for ASCoR contributions, see Brants et al., 2010; Ohme, 2021; Otto et al., 2018), the role of such spirals within the framework of disenchantment with political information remains more elusive. Especially in light of the previous point, further research should investigate whether aggressive and polarised stances in the public are generated by exposure to "dark" political information (or moderate its effectiveness), but also provide incentives to elites to see politics through the lens of partisan conflict who, by relying on negative and uncivil rhetoric, simply "cater to their base"—and so forth. Empirical models able to investigate the unfolding of such reinforcing spirals over time will likely be central to develop more integrated frameworks.

Although research examining negative attitudes towards news has increased considerably over the last few years, it is difficult to grasp the mechanisms behind the different forms of news avoidance and understand the possible consequences for democracy. An additional challenge lies in the measurement of these concepts. Previous studies mainly rely on survey research, while ideal for studying relationships between media exposure and attitudes towards news and politics, they pose measurement issues surrounding self-reported media exposure and attitudes (de Vreese & Neijens, 2016) as well as selection bias, known to specifically affect respondent groups of interest, such as more radical and disaffected citizens. The first can be circumvented with more direct measures of media exposure, specifically online, such as using a browser plug-in, generating insights into media use behaviour (Trilling et al., 2024; Vermeer, et al., 2020). The latter is more difficult to tackle, especially in a context in which these disaffected citizens not only question politics and media, but are increasingly weary of us academics.

To conclude, the findings presented in this chapter outline the need for remedies to counter the nefarious consequences of political information—and negative attitudes towards it. All societal stakeholders likely have an important role to play here. Political actors hold the key to providing a civil, substantive, and democratic alternative to the negative and uncivil campaigns that are so attractive to voters. Recent research suggesting that the decision to go negative on their rivals might come with substantial costs—most notably, voters defecting towards parties that are ideologically close to the attacker (Mendoza et al., in press). This seems to suggest that parties also have strategic reasons to go positive, beyond normative ones.

Policymakers, on their end, have considerable powers to intervene both on the upstream and the downstream mechanisms discussed here. On the one hand, they can support educational and pedagogical initiatives to increase civic interest in political information, boost participation and critical thinking, and provide (future) citizens with sharper tools to "prebunk" any misinformation they might encounter. Especially in light of recent research showing that misinformation effects are harder to curb afterwards (Chan & Albarracín, 2023), taking the lead and developing cognitive skills to recognise it and "inoculate" against its possible nefarious effects seems to be the way to go. On the other hand, policymakers can hold accountable news platforms, including social media, to provide more accurate and "factual" information—a growing challenge, certainly, in light of recent advances in generative artificial intelligence and the complicated legislative task ahead.

Citizens themselves also have a role to play, of course, in being more selective towards "bad" information—not simply towards information they dislike ideologically. While the "negativity bias" at play in all human interactions suggests that "negative" information will always have a comparative advantage over "positive" or "neutral" information, citizens certainly have shown that they can reject excessive forms of political communication—communication that is very uncivil, for instance. Recent research (Valli & Nai, 2023a, 2023b) seems indeed to suggest that citizens can activate "resistance strategies" to curb attempts of political persuasion they are exposed to—for instance, by bolstering their initial attitudes or framing the persuader under a negative light—above and beyond simple mechanisms of motivated reasoning.

News organisations have to find new ways to connect with citizens who have access to an unprecedented amount of news content as well as convince them that paying attention to news is worth their while. Unfortunately, also driven by an increasingly competitive media market across the world, news organisation still face strong incentives to frame information in ways that "sells"—if it bleeds, it leads. Perhaps the most dramatic consequence of this special attention provided by news media to the dark side of politics (Maier & Nai, 2020) is that such focus likely "pours fuel onto the fire" and exacerbates the nefarious effects of exposure to the dark information discussed above.

Finally, researchers and experts increasingly face challenges to communicate their findings effectively to the public. Especially in a context where ideological stances about the topics investigated likely differ within the public—for instance, there is certainly no societal consensus about what constitutes populism, and whether it is a normatively detrimental phenomenon—scientific objectivity and rigor can make the difference between effective and ineffective communication. Rising levels of anti-intellectualism will likely make this task more challenging—and, consequently, even more important—in years to come.

References

Bakker, B. N., Schumacher, G., & Rooduijn, M. (2021). The populist appeal: Personality and antiestablishment communication. *The Journal of Politics, 83*(2), 589–601. https://doi.org/10.1086/710014

Bjarnøe, C., de Vreese, C. H., & Albæk, E. (2020). The effect of being conflict non-avoidant: Linking conflict framing and political participation. *West European Politics, 43*(1), 102–128. https://doi.org/10.1080/01402382.2019.1572334

Boomgaarden, H. G., & Vliegenthart, R. (2007). Explaining the rise of anti-immigrant parties: The role of news media content. *Electoral Studies, 26*(2), 404–417. https://doi.org/10.1016/j.electstud.2006.10.018

Boomgaarden, H. G., & Vliegenthart, R. (2009). How news content influences anti-immigration attitudes: Germany, 1993–2005. *European Journal of Political Research, 48*(4), 516–542. https://doi.org/10.1111/j.1475-6765.2009.01831.x

Bos, L., Egelhofer, J. L., & Lecheler, S. K. (2023). Short but critical? How "fake news" and "anti-elitist" media attacks undermine perceived message credibility on social media. *Communication Research*. Advance online publication. https://doi.org/10.1177/00936502231178432

Bos, L., Kruikemeier, S., & de Vreese, C. (2016). Nation binding: How public service broadcasting mitigates political selective exposure. *PloS ONE, 11*(5), Article e0155112. https://doi.org/10.1371/journal.pone.0155112

Bos, L., Schemer, C., Corbu, N., Hameleers, M., Andreadis, I., Schulz, A., Schmuck, D., Reinemann, C., & Fawzi, N. (2020). The effects of populism as a social identity frame on persuasion and mobilisation: Evidence from a 15-country experiment. *European Journal of Political Research, 59*(1), 3–24. https://doi.org/10.1111/1475-6765.12334

Bos, L., Wichgers, L., & van Spanje, J. (2021). Are populists politically intolerant? Citizens' populist attitudes and tolerance of various political antagonists. *Political Studies, 71*(3), 851–868. https://doi.org/10.1177/00323217211049299

Boukes, M., & Boomgaarden, H. G. (2015). Soft news with hard consequences? Introducing a nuanced measure of soft versus hard news exposure and its relationship with political cynicism. *Communication Research, 42*(5), 701–731. https://doi.org/10.1177/0093650214537520

Brants, K., de Vreese, C., Möller, J., & van Praag, P. (2010). The real spiral of cynicism? Symbiosis and mistrust between politicians and journalists. *The International Journal of Press/Politics, 15*(1), 25–40. https://doi.org/10.1177/1940161209351005

Chan, M. P. S., & Albarracín, D. (2023). A meta-analysis of correction effects in science-relevant misinformation. *Nature Human Behaviour*. Advance online publication. https://doi.org/10.1038/s41562-023-01623-8

Damstra, A., Jacobs, L., Boukes, M., & Vliegenthart, R. (2021). The impact of immigration news on anti-immigrant party support: Unpacking agenda-setting and issue ownership effects over time. *Journal of Elections, Public Opinion and Parties*, *31*(1), 97–118. https://doi.org/10.1080/17457289.2019.1607863

Damstra, A., Vliegenthart, R., Boomgaarden, H., Glüer, K., Lindgren, E., Strömbäck, J., & Tsfati, Y. (2023). Knowledge and the news: An investigation of the relation between news use, news avoidance, and the presence of (mis) beliefs. *The International Journal of Press/Politics*, *28*(1), 29–48. https://doi.org/10.1177/19401612211031457

de Bruin, K., de Haan, Y., Vliegenthart, R., Kruikemeier, S., & Boukes, M. (2021). News avoidance during the COVID-19 crisis: Understanding information overload. *Digital Journalism*, *9*(9), 1286–1302. https://doi.org/10.1080/21670811.2021.1957967

de Vreese, C. H., & Boomgaarden, H. (2006). News, political knowledge and participation: The differential effects of news media exposure on political knowledge and participation. *Acta Politica*, *41*(4), 317–341. https://doi.org/10.1057/palgrave.ap.5500164

de Vreese, C. H., & Neijens, P. (2016). Measuring media exposure in a changing communications environment. *Communication Methods and Measures*, *10*(2–3), 69–80. https://doi.org/10.1080/19312458.2016.1150441

Dobber, T., Metoui, N., Trilling, D., Helberger, N., & de Vreese, C. (2021). Do (micro-targeted) deepfakes have real effects on political attitudes? *The International Journal of Press/Politics*, *26*(1), 69–91. https://doi.org/10.1177/1940161220944364

Elenbaas, M., de Vreese, C., Schuck, A., & Boomgaarden, H. (2014). Reconciling passive and motivated learning: The saturation-conditional impact of media coverage and motivation on political information. *Communication Research*, *41*(4), 481–504. https://doi.org/10.1177/0093650212467032

Hameleers, M. (2019). The populism of online communities: Constructing the boundary between "blameless" people and "culpable" others. *Communication Culture & Critique*, *12*(1), 147–165. https://doi.org/10.1093/ccc/tcz009

Hameleers, M. (2021). *Populist disinformation in fragmented information settings: Understanding the nature and persuasiveness of populist and post-factual communication*. Routledge. https://doi.org/10.4324/9781003194668

Hameleers, M., Bos, L., & de Vreese, C. H. (2017). "They did it": The effects of emotionalized blame attribution in populist communication. *Communication Research*, *44*(6), 870–900. https://doi.org/10.1177/0093650216644026

Hameleers, M., Bos, L., & de Vreese, C. H. (2018). Selective exposure to populist communication: How attitudinal congruence drives the effects of populist attributions of blame. *Journal of Communication*, *68*(1), 51–74. https://doi.org/10.1093/joc/jqx001

Kruikemeier, S., Boerman, S. C., & Bol, N. (2020). Breaching the contract? Using social contract theory to explain individuals' online behavior to safeguard privacy. *Media Psychology, 23*(2), 269–292. https://doi.org/10.1080/15213269.2019.1598434

Maier, J., & Nai, A. (2020). Roaring candidates in the spotlight: Campaign negativity, emotions, and media coverage in 107 national elections. *The International Journal of Press/Politics, 25*(4), 576–606. https://doi.org/10.1177/1940161220919093

Maier, J., Stier, S., & Oschatz, C. (2022). Are candidates rational when it comes to negative campaigning? Empirical evidence from three German candidate surveys. *Party Politics, 29*(4), 766–779. https://doi.org/10.1177/13540688221085239

Mendoza, P., Nai, A., & Bos, L. (In press). The fleeting allure of dark campaigns: Backlash from negative campaigning in presence of (better) alternatives. *Political Communication.*

Möller, J., Kühne, R., & De Vreese, C. (2018). Mobilizing youth in the 21st century: How digital media use fosters civic duty, information efficacy, and political participation. *Journal of Broadcasting & Electronic Media, 62*(3), 445–460. https://doi.org/10.1080/08838151.2018.1451866

Nai, A. (2019). Voter information processing and political decision making. In D. P. Redlawsk (Ed.), *The Oxford encyclopedia of political decision making*. Oxford University Press. https://doi.org/10.1093/acrefore/9780190228637.013.912

Nai, A. (2020). Going negative, worldwide: Towards a general understanding of determinants and targets of negative campaigning. *Government and Opposition, 55*(3), 430–455. https://doi.org/10.1017/gov.2018.32

Nai, A. (2021). Fear and loathing in populist campaigns? Comparing the communication style of populists and non-populists in elections worldwide. *Journal of Political Marketing, 20*(2), 219–250. https://doi.org/10.1080/15377857.2018.1491439

Nai, A., & Maier, J. (2021). Is negative campaigning a matter of taste? Political attacks, incivility, and the moderating role of individual differences. *American Politics Research, 49*(3), 269–281. https://doi.org/10.1177/1532673X20965548

Nai, A., Schemeil, Y., & Marie, J. L. (2017). Anxiety, sophistication, and resistance to persuasion: Evidence from a quasi-experimental survey on global climate change. *Political Psychology, 38*(1), 137–156. https://doi.org/10.1111/pops.12331

Nai, A., Schemeil, Y., & Valli, C. (2023). A persuadable type? Personality traits, dissonant information, and political persuasion. *International Journal of Communication, 17*, 1061–1082. https://ijoc.org/index.php/ijoc/article/view/19182

Nai, A., & Young, E. L. (2023). *Blood on the campaign trail: Negative campaign rhetoric and political violence* [Manuscript submitted for publication]. Amsterdam School of Communication Research, University of Amsterdam.

Neijens, P. C., & de Vreese, C. (2009). Helping citizens decide in referendums: The moderating effect of political sophistication on the use of the Information and

Choice Questionnaire as a decision aid. *Public Opinion Quarterly, 73*(3), 521–536. https://doi.org/10.1093/poq/nfp040

Newman, N., Fletcher, R., Robertson, C. T., Eddy, K., & Nielsen, R. K. (2022). *Reuters Institute digital news report 2022*. Reuters Institute. https://reutersinstitute.politics.ox.ac.uk/sites/default/files/2022-06/Digital_News-Report_2022.pdf

Ohme, J. (2021). Algorithmic social media use and its relationship to attitude reinforcement and issue-specific political participation: The case of the 2015 European immigration movements. *Journal of Information Technology & Politics, 18*(1), 36–54. https://doi.org/10.1080/19331681.2020.1805085

Otto, L. P., Lecheler, S., & Schuck, A. R. (2020). Is context the key? The (non-)differential effects of mediated incivility in three European countries. *Political Communication, 37*(1), 88–107. https://doi.org/10.1080/10584609.2019.1663324

Otto, L., Thomas, F., & Maier, M. (2018). Everyday dynamics of media skepticism and credibility. In K. Otto & A. Köhler (Eds.), *Trust in media and journalism: Empirical perspectives on ethics, norms, impacts and populism in Europe* (pp. 111–133). Springer VS. https://doi.org/10.1007/978-3-658-20765-6_7

Schäfer, S., Aaldering, L., & Lecheler, S. (2022). "Give me a break!" Prevalence and predictors of intentional news avoidance during the COVID-19 pandemic. *Mass Communication and Society, 26*(4), 671–694. https://doi.org/10.1080/15205436.2022.2125406

Schuck, A. R. (2017). Media malaise and political cynicism. In P. Rössler (Ed.), *The international encyclopedia of media effects* (pp. 1–19). Wiley-Blackwell. https://doi.org/10.1002/9781118783764.wbieme0066

Schuck, A. R., Vliegenthart, R., & de Vreese, C. H. (2016). Who's afraid of conflict? The mobilizing effect of conflict framing in campaign news. *British Journal of Political Science, 46*(1), 177–194. https://doi.org/10.1017/S0007123413000525

Simon, M., Welbers, K., Kroon, A. C., & Trilling, D. (2022). Linked in the dark: A network approach to understanding information flows within the Dutch Telegramsphere. *Information, Communication & Society*. Advance online publication. https://doi.org/10.1080/1369118X.2022.2133549

Stier, S., Kirkizh, N., Froio, C., & Schroeder, R. (2020). Populist attitudes and selective exposure to online news: A cross-country analysis combining web tracking and surveys. *The International Journal of Press/Politics, 25*(3), 426–446. https://doi.org/10.1177/1940161220907018

Trilling, D., Araujo, T., Kroon, A., Möller, A. M., Strycharz, J., & Vermeer, S. (2024). Computational communication science in a digital society. In T. Araujo & P. Neijens (Eds.), *Communication research into the digital society: Fundamental insights from the Amsterdam School of Communication Research* (pp. 247–263). Amsterdam University Press. https://doi.org/10.5117/9789048560592_ch15

Valli, C., & Nai, A. (2023a). Dispositioned to resist? The Big Five and resistance to dissonant political views. *Personality and Individual Differences*, *207*, Article 112152. https://doi.org/10.1016/j.paid.2023.112152

Valli, C., & Nai, A. (2023b). Let me think about it: Cognitive elaboration and strategies of resistance to political persuasion. *Media Psychology*, *26*(2), 89–112 https://doi.org/10.1080/15213269.2022.2098774

van der Meer, T. G., Hameleers, M., & Ohme, J. (2023). Can fighting misinformation have a negative spillover effect? How warnings for the threat of misinformation can decrease general news credibility. *Journalism Studies*, *24*(6), 803–823. https://doi.org/10.1080/1461670X.2023.2187652

van Reijmersdal, E. A., Fransen, M. L., van Noort, G., Opree, S. J., Vandeberg, L., Reusch, S., van Lieshout, F., & Boerman, S. C. (2016). Effects of disclosing sponsored content in blogs: How the use of resistance strategies mediates effects on persuasion. *American Behavioral Scientist*, *60*(12), 1458–1474. https://doi.org/10.1177/0002764216660141

Vermeer, S., Kruikemeier, S., Trilling, D., & de Vreese, C. (2022). Using panel data to study political interest, news media trust, and news media use in the early stages of the COVID-19 pandemic. *Journalism Studies*, *23*(5–6), 740–760. https://doi.org/10.1080/1461670X.2021.2017790

Vermeer, S., Trilling, D., Kruikemeier, S., & de Vreese, C. (2020). Online news user journeys: The role of social media, news websites, and topics. *Digital Journalism*, *8*(9), 1114–1141. https://doi.org/10.1080/21670811.2020.1767509

Walter, A. S., & Vliegenthart, R. (2010). Negative campaigning across different communication channels: Different ball games? *The International Journal of Press/Politics*, *15*(4), 441–461. https://doi.org/10.1177/1940161210374122

Zarouali, B., Strycharz, J., Helberger, N., & de Vreese, C. (2022). Exploring people's perceptions and support of data-driven technology in times of COVID-19: The role of trust, risk, and privacy concerns. *Behaviour & Information Technology*, *41*(10), 2049–2060. https://doi.org/10.1080/0144929X.2021.2022208

About the authors

Alessandro Nai is an Associate Professor of Political Communication & Journalism at ASCoR. His research focuses on the dark sides of politics, the use of negativity and incivility in election campaigns in a comparative perspective, and the (dark) personality of political figures. E-mail: A.Nai@uva.nl

Susan Vermeer was an Assistant Professor at ASCoR and currently holds the same title at Wageningen University & Research. By combining traditional research methods with innovative and computational methods, she aims to understand the impact of the digital media environment, such as social media and algorithms, on political behaviour and attitudes. E-mail: Susan1.Vermeer@wur.nl.

Linda Bos is an Associate Professor and Programme Group Director of Political Communication & Journalism at ASCoR. She is an expert in the field of populist political communication. Her broader line of research centres on the relation between political elites, media, and voters. E-mail: L.Bos@uva.nl

Michael Hameleers is an Assistant Professor of Political Communication & Journalism at ASCoR. His research interests include (right-wing) populism, disinformation, and selective exposure. He has published extensively on (visual) disinformation and the effects of populism. E-mail: M.Hameleers@uva.nl

5. Youth and the Digital Society

Jessica Taylor Piotrowski

Abstract
The digital transformation has impacted all aspects of society. These impacts are most acutely seen amongst the youngest generation as they have been born into a world that looks entirely different than generations before. For researchers interested in how youth navigate, accommodate, and shape the digital society, these changes have precipitated a need for revised theorising; more precise analytic approaches; and a recognition that many young people lack the skills necessary to fully participate in this digital society. ASCoR scholars have contributed meaningfully to a global dialogue about how we can ensure that youth are prepared for the digital world they live in and remain committed to this dynamic dialogue in the years ahead.

Keywords: adolescence, children, differential susceptibility, technological access, digital competence

Introduction

After the introduction of television, E. B. White famously said that television was "going to be the test of the modern world. We shall stand or fall by the television—of that I am quite sure" (White, 1950). Fast forward to 2023—and replace television with digitisation. Every aspect of our mediated world is, or can be, digital. Access is at one's fingertips, information should be bite-sized, experiences should be interactive, narrative should be transportive, and communication can be through and with human and non-human agents. The digital transformation has impacted all aspects of society. These impacts are perhaps most acutely seen amongst the youngest generation as they have been born into a world that looks entirely different than generations

Piotrowski, J. T. (2024). Youth and the digital society. In T. Araujo & P. Neijens (Eds.), *Communication research into the digital society: Fundamental insights from the Amsterdam School of Communication Research* (pp. 87-100). Amsterdam University Press. https://doi.org/10.5117/9789048560592_ch05

before. Communication scholars have always identified this group as a special one for study, but digital transformations have placed this group in a space of urgency—with society demanding answers regarding the impact of social media on youth; the extent to which (emerging) technology such as smartphones, wearables, streaming services, virtual assistants, social robots, and artificial intelligence might reshape childhood; the ethical ramifications of the digital society (see Livingstone & Third, 2017), including the right to participation and representation, the right to (data) protection and online safety, the right to digital competence, and the right to be forgotten; and the skills necessary to thrive in this new digital space.

In the past decade, youth and media scholars at ASCoR have worked to respond to these changing times and these pressing issues. Given that the digital society has impacted many different echelons of youth and media scholarship, there is not one theoretical, methodological, or conceptual development that can be highlighted as pinnacle. Rather, there are a handful of interconnected developments that together have shaped—and continue to shape—how ASCoR researchers approach and contribute to the dynamic topic of youth and media in the digital society.

Theoretical contributions

Perhaps most impactful to the youth and media field has been Valkenburg and Peter's (2013) theoretical publication "The Differential Susceptibility to Media Effects Model." After only a decade this publication has already been cited nearly a thousand times and is recognised by scholars globally as game-changing—both for communication science and for youth and media scholars specifically. In this manuscript, Valkenburg and Peter argue for attention to differential susceptibility—namely, a recognition that not all media effect all users in the same manner. To that end, they offer a theoretical model, DSMM, which posits that individual differences in developmental, dispositional, and social susceptibility impact media selection and media processing—leading to differential effects.

For certainty, even before the digital transformation, the communication science field saw us theorising about differential effects to some extent. But Valkenburg and Peter's model, whose development not coincidently occurred as digital technology was fiercely grabbing hold of the youth media landscape, has pushed scholars to rethink existing practices. In the analogue media era, most (youth and media) scholars relied on statistical controls to draw conclusions for all (Piotrowski & Valkenburg, 2015). As a result, differential susceptibility was often excluded entirely from analytic considerations, and instead the tradition leaned towards either saying, "Yes, effect for all" or "No, no effect for all." But the digital transformations in the field—moving, for example, from analogue television, books, and radio to social media, games, interactive television, social robots, virtual assistants, and (personal) smart devices—made it clear that the field needed to consider "Yes, for some" as a plausible (and likely) outcome.

Consider social media, such as TikTok or YouTube, for a moment. The uptick of social media use among youth (who are among the earliest adopters and most frequent users of social media across all age groups) has been a field-changer for youth and media scholars. Youth today choose amongst an array of social media spaces where they then co-construct a unique experience that is all their own. Both TikTok and YouTube offer a video viewing experience, both offer opportunities for short form video content, and both offer opportunities for social comments. Yet, they offer a very different experience when compared to each other. Even more, within TikTok or YouTube, teens will have a very different experience depending upon the choices they make. Inasmuch, simply offering a "Yes" or a "No" as to whether social media impacts young people is illogical. We need to understand which youth choose which social media, how they use this social media, what responses they experience when using this social media, and what effects occur thereafter.

The DSMM is the first comprehensive theory to provide theoretical space for these questions, and ASCoR scholars gave the DSMM its first voice via a pair of studies. First, in a longitudinal study with adolescents, ASCoR scholars were interested in nuancing the debate regarding (digital and analogue) media violence and aggressive behaviour. Rather than "Yes" or "No," they found "Yes, for some." Specifically, they demonstrated that children growing up in families with higher conflict are more susceptible to the impact of media violence as these teens experience greater arousal when consuming violent media content and, as a result, demonstrate increased aggressive tendencies (Fikkers, 2016; Fikkers et al., 2013; Fikkers et al., 2016). Similarly, in a parallel longitudinal study with younger children, ASCoR researchers asked about the relationship between ADHD behaviour and violent/scary media. Here, too, a "Yes, for some" relationship was found. Boys with increased ADHD behaviour were found to consume more violent/scary media and demonstrated differences in arousal and attention to violent/scary media. Moreover, the use of autonomy-supportive parenting was associated with fewer ADHD behaviours and less violent/scary media consumption, and lastly, ADHD behaviours longitudinally predicted violent/scary media content consumption which then increased ADHD behaviours (Nikkelen et al., 2014; Nikkelen et al., 2015).

These two case studies were foundational examples of the utility of the DSMM in the digital society. And, since that time, the use of the model across the field has exploded—paralleling the explosion of the use of digital media throughout childhood and adolescence. ASCoR scholars have been at the forefront of this explosion—asking precisely when, for whom, and how the digital society impacts young people. This work has touched upon multitasking, virtual assistants, sexual internet content, and more. For example, Baumgartner and colleagues found longitudinal evidence for a potential detrimental long-term effect of media multitasking on attention problems for early adolescents (Baumgartner et al., 2017; see also van der Schuur et al., 2015). Van Oosten (2016) demonstrated that the link between sexually explicit internet use and sexual uncertainty was only present among girls with low hyper-gendered orientation and a high impersonal sex orientation. Wald and colleagues (Wald, Piotrowski, van Oosten, et al., 2023) found that differences in technology confidence, internet literacy, and preferred style of media mediation best characterise whether families have (34%) smart speakers in their homes. And Meier et al. (2022) showed that while more automatic social media use and more frequent phone checking predicted procrastination for teens, this occurred only in a minority of adolescents.

Methodological contributions

Just as the theorising has shifted to encourage more nuanced answers about today's digital society, this theorising—combined with a quickly changing media landscape—has necessitated a shift in methods as well. The digital society affords (and often assumes) that the user is an active contributor in the experience—engaging via and with technology. Children and teens engage with digital games (Lemmens et al., 2015); they text with friends (Valkenburg & Peter, 2009); they post images on Instagram and comments on YouTube (de Vries et al., 2018; Möller et al., 2021); they play with apps (Broekman et al., 2018); they talk to virtual assistants (Wald, Piotrowski, Araujo, et al., 2023); they high-five robots (Peter & Kühne, 2018); and they are surrounded by others using digital media throughout their everyday lives (Wolfers et al., 2020). Youth are among the first adopters of changing technologies, and these technologies have left us asking: What are the differential impacts? and How do we measure the impact? In a space where simply posting a status update can impact one's own sense of self (Valkenburg, 2017), the digital society has required scholars to reconsider our measurement and analytic approaches.

In this book, Trilling and colleagues (2024) highlight the so-called computational turn at ASCoR—demonstrating how the digital society has pushed ASCoR scholars to leverage computational methodology to answer key questions of our time. This computational turn has also been felt in the youth and media sector, with scholars relying on computational methodology (via public data scraping and personal data donation) to assess large corpuses of (media) content. For example, computational approaches have been used to assess the types of sexual information that adolescents locate online; total amount of smartphone use (Baumgartner et al., 2023); and the types of behaviours engaged in during social media usage (van Driel et al., 2022). These approaches have been a valuable addition in an age of big data, handheld devices, and personalised experiences. They help us understand aspects of the so-called black box (Fikkers & Piotrowski, 2020)—namely, what individuals are consuming as well as where and how they are responding to this information (see also Araujo et al., 2022). Yet, while these computational methods provide us the opportunity to gather more precise insight into the media experience, they do not by default capitalise on this precision—leaving incomplete answers to key questions in the field.

Specifically, the media effects field—within and outside youth and media—has long suffered with a mismatch between theorising and analysis. Theoretically, the media effects tradition acknowledges that media effects occur within a person and that the type of effects depend upon the interaction between person, content, and context characteristics (i.e., DSMM predictions). Yet historically, analyses have primarily looked for differences between persons. Indeed, a look across the youth and media space (e.g., Valkenburg & Piotrowski, 2017) shows that the great majority of (quantitative) studies have either compared one group with another group via (quasi-)experiments or relied on a single (occasionally ±3 data points) for a single participant in survey/longitudinal research. With such relatively limited datasets, it was only possible to statistically compare—and make conclusions—between people. And for the most part, this was sufficient, particularly with analogue media where the content was largely known, limited in scope, and interactivity was non-existent.

The digital society, however, brings with it nearly limitless media experiences that are often precipitated upon the types of interactivities involved. The increased accessibility of computational methods and in-depth measurement (e.g., experience sampling) (see Siebers et al., 2022; Verbeij et al., 2022) offers researchers the statistical and computational power to zoom into an individual participant's data and actually uncover within-person relationships between predictors and outcomes. Meier's work, for example, capitalised on 22,809 assessments from 312 adolescents (Meier et al., 2022). With such data, it is increasingly possible to study the very relationships that scientists have been theorising about all along; enabling scholars to augment existing work by looking within (many) persons and, in doing so, obtaining richer detail about for whom these relationships exist or not.

ASCoR scholars have responded to this opportunity by encouraging the youth and media field to consider the use of person-specific approaches (Valkenburg et al., 2021) to precisely understand what the relationship between youth and the digital society is, and how the digital society is (re) shaping childhood and adolescence. For example, ASCoR scholars in Project AWeSome implemented a person-specific analytic paradigm to test how each young person is impacted. Rather than concluding a null relationship, which would have been the case with former approaches, they found that the majority of adolescents (88%) experience little to no effect of social media use on self-esteem but 4% experienced positive effects and 8% experienced negative effects (Valkenburg et al., 2021). At a societal level, this is a meaningful distinction. And, indeed, similar patterns have been echoed with analyses on well-being with 45% of adolescents experiencing no changes in well-being due to social media use, 28% experiencing declines in well-being, and 26% experiencing increases in well-being (Beyens et al., 2021). This work, and others like it from ASCoR scholars, has helped shift the dialogue about effects in the digital society from "Is there an effect?" to "For whom is there which effect?"—making a strong argument for including more advanced person-specific approaches in digital society scholarship.

Conceptual developments

While theorising and methodological approaches have evolved alongside changes in the digital society, so have the conversations about what it means for youth to thrive in the digital society. Here, too, ASCoR scholars have been actively engaging in scholarship to understand the protective and empowering factors that may be unique to this space. One topic that has entered the lexicon of many ASCoR scholars is digital competence—namely, the digital knowledge and digital skills needed (for youth) to thrive in the digital environment.

Note that the phrase "for youth" is indicated parenthetically as the concept is truly a lifespan concept. The digital society is a powerfully beneficial force for some individuals in some contexts. But these benefits are neither uniform nor equitable. There are groups who remain excluded or marginalised because of a lack of access to technology transformations. And, even within a prosperous country such as the Netherlands, data collected in ASCoR's Digital Competence (DIGCOM) Project shows that technology access is not equal (e.g., de Vries et al., 2022b). At the same time, nearly half of the European population lacks the knowledge and skills necessary to thrive in the digital world (Clifford et al., 2020; European Union, n.d.). Without the right skills, the opportunities of the digital society are unlikely to be experienced. Instead, the pitfalls—privacy and security leaks; mental health and well-being degradation; digital addiction, and more—risk becoming the hallmark of the digital society.

This is an urgent problem, which will only be exacerbated as artificial intelligence advances at rapid speed. Solving the problem requires intervention at numerous societal levels, with youth considered a critical point of impact. In the Netherlands, for example, there is a robust dialogue about the degree to which digital competence should be a required learning outcome in early education. But enacting such a policy means understanding precisely what digital competence is, and what knowledge and capacities youth do (and do not) have. Only then is it possible to investigate digital diversity: namely, who requires support, what type of support is needed, and how best to offer it (de Vries et al., 2023).

As first step, ASCoR researchers have contributed to this dialogue by creating the DigIQ®—a psychometrically valid assessment tool to assess digital competence that covers the full dimension of digital competence (strategic information; critical information; netiquette; digital content creation; safety and control of information and devices; digital health and well-being; sustainable/green technology; artificial intelligence), captures variability, and facilitates comparisons across age (de Vries et al., 2022a). Even more, this tool provides connections to local resources to help individuals bolster their skills and is now part of a national dialogue about ways to monitor and support digital competence. By formalising and elevating the concept of digital competence to the (inter)national agenda, ASCoR researchers have offered a critical foundation for future dialogue on thriving in the digital society.

Next steps

At ASCoR, this foundational conversation of digital competence is already being complemented by research on growing up in a digital society. From asking how and whether children form relationships with social robots (de Jong et al., 2019; Peter et al., 2019; Peter & Kühne, 2018; van Straten et al., 2022), to asking about their interaction and accommodation of virtual assistants (Wald, Piotrowski, Araujo, et al., 2023), to questions about youth multitasking online (Baumgartner et al., 2017), to studying the degree to which digital spaces support healthy solitude (Keessen et al., 2022)—ASCoR researchers continue to make clear that young people are confronted with a world that their (grand)parents never experienced.

It will take all of us—including the developers of this digital space (as argued by Dekker et al., 2023)—to ensure the road ahead does more good than harm. This means pushing for a responsible tech agenda, including ethics by design (e.g., Helberger et al., 2018; Palomar-Garcia et al., 2023; Slater et al., 2020). Too long have developers taken an agnostic approach to the effects of their technology—advancing the position that they are not responsible for how their technology is used. Yet, they are also responsible. They are in possession of a wealth of data that informs them about how users respond to their content—including how "like" features, stopping cues, editing features, time indicators, data privacy tools, and more influence the experience and safety of users. Developers can use this data to improve the media ecosystem, and in doing so, benefit all of us. It is a choice to allow algorithmic recommendations to reflect and reinforce inequalities in our social system (Helberger et al., 2018). It is a choice to prioritise and push persuasive influencer content to teens (van Reijmersdal & van Dam, 2020). It is a choice to create platforms that are addictive (Lemmens & Hendriks, 2016). The future involves holding developers accountable, too. Developers, researchers, policymakers, and youth themselves—all need to sit at the same table. ASCoR researchers are ready to play their part. Just recently, for example, Sindy Sumter, Chei Billedo, and Irene van Driel launched SeeMeBeMe—an initiative in which ASCoR researchers are engaging with youth, developers, and policymakers to share research insights on how youth experience media to better ensure that "future media products … support ALL young people in today's digital, hyperdiverse society" (SeeMeBeMe, 2023).

At the same time, the future also involves concretely operationalising the in vogue term "digital well-being." At the time of this writing, it functions as an omnibus term for any positive integration of technology into one's life—and, in that regard, there is real opportunity to refine and nuance this concept. A cursory review of the literature suggests that digital well-being refers to the state of one's physical, mental, and emotional health in relation to one's use of technology and digital devices (e.g., Burr et al., 2020; Orben & Przybylski, 2019; Vanden Abeele, 2021). It is said to involve finding a balance between the benefits and risks associated with digital technology and is often connected to practices such as setting boundaries for screen time, taking technology breaks, using technology in a responsible way, and developing healthy habits around technology use. While on its face this is an interesting concept, there is much to be gained in specifying this concept and its sub-concepts—both for measurement and intervention.

Lastly, and perhaps most difficult, the future requires changing the field's current reactive stance to a proactive one whereby scholarship asks how we can ensure that youth are prepared for the next digital development. Historically, communication science has demonstrated that—with each new technology—society first experiences fear and panic, with the so-called few "innovators" excitedly trying the technology (Livingstone, 2002). This is often followed by resistance and scepticism, as people question the value added. With time, integration often occurs—and then discussions about limits and boundaries emerge. But as history has shown, particularly as it relates to ethics and privacy, these discussions are often too late. These discussions need to be at the forefront so that the process of technological accommodation keeps humans in the loop throughout the process, not only after.

References

Araujo, T., Ausloos, J., van Atteveldt, W., Loecherbach, F., Moeller, J., Ohme, J., Trilling, D., van de Velde, B., de Vreese, C., & Welbers, K. (2022). OSD2F: An open-source data donation framework. *Computational Communication Research, 4*(2), 372–387. https://doi.org/10.5117/CCR2022.2.001.ARAU

Baumgartner, S. E., Sumter, S. R., Petkevič, V., & Wiradhany, W. (2023). A novel iOS data donation approach: Automatic processing, compliance, and reactivity in a longitudinal study. *Social Science Computer Review, 41*(4), 1456–1472. https://doi.org/10.1177/08944393211071068

Baumgartner, S. E., van der Schuur, W. A., Lemmens, J. S., & te Poel, F. (2017). The relationship between media multitasking and attention problems in adolescents: Results of two longitudinal studies. *Human Communication Research, 44*(1), 3–30. https://doi.org/10/gfw9jg

Beyens, I., Pouwels, J. L., van Driel, I. I., Keijsers, L., & Valkenburg, P. M. (2021). Social media use and adolescents' well-being: Developing a typology of person-specific effect patterns. *Communication Research*. Advance online publication. https://doi.org/10.1177/00936502211038196

Broekman, F. L., Piotrowski, J. T., Beentjes, H. W., & Valkenburg, P. M. (2018). App features that fulfill parents' needs in apps for children. *Mobile Media & Communication, 6*(3), 367–389. https://doi.org/10.1177/2050157918759571

Burr, C., Taddeo, M., & Floridi, L. (2020). The ethics of digital well-being: A thematic review. *Science and Engineering Ethics, 26*(4), 2313–2343. https://doi.org/10.1007/s11948-020-00175-8

Clifford, I., Kluzer, S., Troia, S., Jakobsome, S., & Zandbergs, U. (2020). *DigCompSat: A self-reflection tool for the European digital framework for citizens*. Publications Office of the European Union. https://doi.org/10.2760/77437

de Jong, C., Kühne, R., Peter, J., van Straten, C. L., & Barco, A. (2019). What do children want from a social robot? Toward gratifications measures for child–robot interaction. *Proceedings of the 2019 28th IEEE International Conference on Robot and Human Interactive Communication (RO-MAN), India*, 1–8. https://doi.org/10.1109/RO-MAN46459.2019.8956319

Dekker, C., A., Baumgartner, S. E., Sumter, S. R., & Ohme, J. (2023, September 20–23). *Improving digital well-being: Findings from two intervention studies targeting smartphone behavior* [Poster session]. 2023 Digital Media and Developing Minds International Scientific Congress, Washington, DC.

de Vries, D. A., Möller, A. M., Wieringa, M. S., Eigenraam, A. W., & Hamelink, K. (2018). Social comparison as the thief of joy: Emotional consequences of viewing strangers' Instagram posts. *Media Psychology, 21*(2), 222–245. https://doi.org/10.1080/15213269.2016.1267647

de Vries, D. A., Piotrowski, J. T., & de Vreese, C. (2022a, December 8). *DigIQ*. OSF. https://osf.io/dfvqb

de Vries, D. A., Piotrowski, J. T., & de Vreese, C. (2022b). *Resultaten onderzoek digitale competenties (DIGCOM)*. University of Amsterdam. https://osf.io/7sn6q/

de Vries, D. A., Piotrowski, J. T., & de Vreese, C. (2023). *Developing DigIQ: A measure of digital competence* [Manuscript in preparation]. Amsterdam School of Communication Research, University of Amsterdam.

European Union. (n.d.). *Digital competence framework for educators (DigCompEdu)*. https://ec.europa.eu/jrc/en/digcompedu

Fikkers, K. M. (2016). *A different(ial) perspective: How social context influences the media violence–aggression relationship among early adolescents* (Publication no. 9789462039681) [Doctoral dissertation, University of Amsterdam]. UvA-DARE. https://hdl.handle.net/11245/1.512223

Fikkers, K. M., & Piotrowski, J. T. (2020). Content and person effects in media research: Studying differences in cognitive, emotional, and arousal responses to media content. *Media Psychology, 23*(4), 493–520. https://doi.org/10/ggkz53

Fikkers, K. M., Piotrowski, J. T., & Valkenburg, P. M. (2016). Beyond the lab: Investigating early adolescents' cognitive, emotional, and arousal responses to violent games. *Computers in Human Behavior, 60*, 542–549. https://doi.org/10.1016/j.chb.2016.02.055

Fikkers, K. M., Piotrowski, J. T., Weeda, W. D., Vossen, H. G. M., & Valkenburg, P. M. (2013). Double dose: High family conflict enhances the effect of media violence exposure on adolescents' aggression. *Societies, 3*(3), 280–292. https://doi.org/10/gcftnk

Helberger, N., Karppinen, K., & D'Acunto, L. (2018). Exposure diversity as a design principle for recommender systems. *Information, Communication & Society, 21*(2), 191–207. https://doi.org/10/gfkn39

Keessen, L., Piotrowski, J. T., Sumter, S. R., & van Oosten, J. M. F. (2022, June 19–23). Me-time in the digital age: Development and validation of the need for solitude scale [Paper presentation]. 26th Biennial Meeting of the International Society for the Study of Behavioural Development, Rhodes, Greece.

Lemmens, J. S., & Hendriks, S. J. F. (2016). Addictive online games: Examining the relationship between game genres and internet gaming disorder. *Cyberpsychology, Behavior, and Social Networking, 19*(4), 270–276. https://doi.org/10.1089/cyber.2015.0415

Lemmens, J. S., Valkenburg, P. M., & Gentile, D. A. (2015). The internet gaming disorder scale. *Psychological Assessment, 27*(2), 567–582. https://doi.org/10/f7gfs4

Livingstone, S. (2002). *Young people and new media: Childhood and the changing media environment.* Sage. https://doi.org/10.4135/9781446219522

Livingstone, S., & Third, A. (2017). Children and young people's rights in the digital age: An emerging agenda. *New Media & Society, 19*(5), 657–670. https://doi.org/10.1177/1461444816686318

Meier, A., Beyens, I., Siebers, T., Pouwels, J. L., & Valkenburg, P. M. (2022). Habitual social media and smartphone use are linked to procrastination for some, but not all, adolescents. PsyArXiv. https://doi.org/10.31234/osf.io/y9fuv

Möller, A. M., Baumgartner, S. E., Kühne, R., & Peter, J. (2021). Sharing the fun? How social information affects viewers' video enjoyment and video evaluations. *Human Communication Research, 47*(1), 25–48. https://doi.org/10.1093/hcr/hqaa013

Nikkelen, S. W. C., Vossen, H. G., & Valkenburg, P. M. (2015). Children's television viewing and ADHD-related behaviors: Evidence from the Netherlands. *Journal of Children and Media, 9*(4), 399–418. https://doi.org/10.1080/17482798.2015.1088872

Nikkelen, S. W. C., Vossen, H. G. M., Valkenburg, P. M., Velders, F. P., Windhorst, D. A., Jaddoe, V. W. V., Hofman, A., Verhulst, F. C., & Tiemeier, H. (2014). Media violence and children's ADHD-related behaviors: A genetic susceptibility perspective. *Journal of Communication*, *64*(1), 42–60. https://doi.org/10/f57fkp

Orben, A., & Przybylski, A. K. (2019). The association between adolescent well-being and digital technology use. *Nature Human Behaviour*, *3*, 173–182. https://doi.org/10.1038/s41562-018-0506-1

Palomar-Garcia, C., Fernández-Planells, A., & Sora-Domenjó, C. (2023). Uses and abuses of User-Centered Design: Literature review and proposal of a tool for evaluating ethics in the design of mobile applications. *Digital Creativity*, *34*(1), 37–52. https://doi.org/10.1080/14626268.2023.2165107

Peter, J., & Kühne, R. (2018). The new frontier in communication research: Why we should study social robots. *Media and Communication*, *6*(3), 73–76. https://doi.org/10/ggkjf3

Peter, J., Kühne, R., Barco, A., de Jong, C., & van Straten, C. L. (2019). Asking today the crucial questions of tomorrow: Social robots and the internet of toys. In G. Mascheroni & D. Holloway (Eds.), *The internet of toys: Practices, affordances and the political economy of children's smart play* (pp. 25–46). Palgrave Macmillan. https://doi.org/10.1007/978-3-030-10898-4_2

Piotrowski, J. T., & Valkenburg, P. M. (2015). Finding orchids in a field of dandelions: Understanding children's differential susceptibility to media effects. *American Behavioral Scientist*, *59*(14), 1776–1789. https://doi.org/10/ggh73t

SeeMeBeMe (2023). Vision. https://seemebeme.org/about-2/

Siebers, T., Beyens, I., Pouwels, J. L., & Valkenburg, P. M. (2022). Explaining variation in adolescents' social media-related distraction: The role of social connectivity and disconnectivity factors. *Current Psychology*. Advance online publication. https://doi.org/10.1007/s12144-022-03844-y

Slater, M., Gonzalez-Liencres, C., Haggard, P., Vinkers, C., Gregory-Clarke, R., Jelley, S., Watson, Z., Breen, G., Schwarz, R., Steptoe, W., Szostak, D., Halan, S., Fox, D., & Silver, J. (2020). The ethics of realism in virtual and augmented reality. *Frontiers in Virtual Reality*, *1*, Article 1. https://doi.org/10.3389/frvir.2020.00001

Trilling, D., Araujo, T., Kroon, A., Möller, A. M., Strycharz, J., & Vermeer, S. (2024). Computational communication science in a digital society. In T. Araujo & P. Neijens (Eds.), *Communication research into the digital society: Fundamental insights from the Amsterdam School of Communication Research* (pp. 247-263). Amsterdam University Press. https://doi.org/10.5117/9789048560592_ch15

Valkenburg, P. M. (2017). Understanding self-effects in social media: Self-effects in social media. *Human Communication Research*, 43(4), 477–490. https://doi.org/10/gcmcjm

Valkenburg, P. M., Beyens, I., Pouwels, J. L., van Driel, I. I., & Keijsers, L. (2021). Social media use and adolescents' self-esteem: Heading for a person-specific media effects paradigm. *Journal of Communication, 71*(1), 56–78. https://doi.org/10.1093/joc/jqaa039

Valkenburg, P. M., & Peter, J. (2009). Social consequences of the internet for adolescents: A decade of research. *Current Directions in Psychological Science, 18*(1), 1–5. https://doi.org/10/cwrkk7

Valkenburg, P. M., & Peter, J. (2013). The differential susceptibility to media effects model. *Journal of Communication, 63*(2), 221–243. https://doi.org/10.1111/jcom.12024

Valkenburg, P. M., & Piotrowski, J. T. (2017). *Plugged in: How media attract and affect youth.* Yale University Press.

Vanden Abeele, M. M. P. (2021). Digital wellbeing as a dynamic construct. *Communication Theory, 31*(4), 932–955. https://doi.org/10.1093/ct/qtaa024

van der Schuur, W. A., Baumgartner, S. E., Sumter, S. R., & Valkenburg, P. M. (2015). The consequences of media multitasking for youth: A review. *Computers in Human Behavior, 53*, 204–215. https://doi.org/10.1016/j.chb.2015.06.035

van Driel, I. I., Giachanou, A., Pouwels, J. L., Boeschoten, L., Beyens, I., & Valkenburg, P. M. (2022). Promises and pitfalls of social media data donations. *Communication Methods and Measures, 16*(4), 266–282. https://doi.org/10.1080/19312458.2022.2109608

van Oosten, J. M. F. (2016). Sexually explicit internet material and adolescents' sexual uncertainty: The role of disposition-content congruency. *Archives of Sexual Behavior, 45*(4), 1011–1022. https://doi.org/10/f65z36

van Reijmersdal, E. A., & van Dam, S. (2020). How age and disclosures of sponsored influencer videos affect adolescents' knowledge of persuasion and persuasion. *Journal of Youth and Adolescence, 49*(7), 1531–1544. https://doi.org/10.1007/s10964-019-01191-z

van Straten, C. L., Peter, J., Kühne, R., & Barco, A. (2022). On sharing and caring: Investigating the effects of a robot's self-disclosure and question-asking on children's robot perceptions and child–robot relationship formation. *Computers in Human Behavior, 129*, Article 107135. https://doi.org/10.1016/j.chb.2021.107135

Verbeij, T., Pouwels, J. L., Beyens, I., & Valkenburg, P. M. (2022). Experience sampling self-reports of social media use have comparable predictive validity to digital trace measures. *Scientific Reports, 12*, Article 7611. https://doi.org/10.1038/s41598-022-11510-3

Wald, R., Piotrowski, J. T., Araujo, T., & van Oosten, J. M. F. (2023). Virtual assistants in the family home. Understanding parents' motivations to use virtual assistants with their child(dren). *Computers in Human Behavior, 139*, Article 107526. https://doi.org/10.1016/j.chb.2022.107526

Wald, R., Piotrowski, J. T., van Oosten, J. M. F., & Araujo, T. (2023). Who are the (non-) adopters of smart speakers? A cross-sectional survey study of Dutch families. *Tijdschrift voor Communicatiewetenschap*. Advance online publication. https://doi.org/10.5117/TCW2023.X.001.WALD

White, E. B. (1950). *One man's meat*. Harper and Brothers.

Wolfers, L. N., Kitzmann, S., Sauer, S., & Sommer, N. (2020). Phone use while parenting: An observational study to assess the association of maternal sensitivity and smartphone use in a playground setting. *Computers in Human Behavior, 102*, 31–38. https://doi.org/10.1016/j.chb.2019.08.013

About the author

Jessica Taylor Piotrowski is a Full Professor at ASCoR, where she holds the Chair in Communication in the Digital Society. As a media psychologist, she focuses on identifying risk, resiliency, and enhancement factors that allow youth to become engaged digital citizens. E-mail: J.Piotrowski@uva.nl

6. The Media Entertainment Success Cycle

Jeroen S. Lemmens

Abstract

The Media Entertainment Success Cycle (MESC) provides a comprehensive theoretical framework that can help researchers identify and understand the supply and demand processes that govern successful entertainment media. The supply process encompasses entertainment development and distribution, aimed at providing an immersive and engaging experience. The demand process illustrates how individual responses and effects influence preference and motivation for the selection of entertainment products. Positive reception not only encourages increased user engagement with similar content but also stimulates the development of similar products within the entertainment industry. The MESC summarises entertainment research within an integrative framework where industry supply and user demand are two mutually reinforcing processes that perpetuate successful media entertainment.

Keywords: industry practices, entertainment development, entertainment distribution, entertainment responses, entertainment preferences

Introduction

Over the last 25 years, entertainment has become a fundamental part of our daily media consumption. The global entertainment industry is enormous and has permeated all layers of communication by transcending numerous forms of media and messages. News outlets, commercial companies, political parties, educational institutions, charities, and religious organisations all employ entertainment elements and strategies to strengthen the appeal of their message, or to enhance its effectiveness. Current definitions of entertainment have moved beyond the traditional notion that entertainment

Lemmens, J. S. (2024). The media entertainment success cycle. In T. Araujo & P. Neijens (Eds.), *Communication research into the digital society: Fundamental insights from the Amsterdam School of Communication Research* (pp. 101-120). Amsterdam University Press. https://doi.org/10.5117/9789048560592_ch06

offers exclusively pleasurable experiences that arise from hedonic content, to include more complex emotional and cognitive experiences (Riddle et al., 2022). Whether through enjoyment, appreciation, contemplation, or wonder, entertainment provides experiences that engage and captivate audiences.

The compelling and effective potential of entertainment is recognised throughout ASCoR. In the context of journalism, people show a strong preference for entertainment news over other news topics, causing them to predominantly select and consume entertainment in online news environments (Vermeer et al., 2020). Research within political communication has shown that satirical programmes on television can either positively or negatively affect the attitude toward a satirised political subject (Boukes et al., 2015). Within persuasion, it is commonly understood that entertaining elements are less likely to activate defence mechanisms compared to overt advertising messages because they facilitate the experience of transportation and identification, while reducing counter-arguing (Asbeek Brusse et al., 2015). Within a health communication perspective, entertainment narratives can discourage binge drinking by reducing associated positive beliefs and attitudes (van Leeuwen et al., 2017). Lastly, related to marketing communications, the overall appeal of brands can be increased through advergames, as these enhance consumers' affective, conative, and behavioural brand outcomes (van Berlo et al., 2021). Yet despite abundant empirical evidence for its global pervasiveness, commercial appeal, societal relevance, and wide-ranging effectiveness, there is no comprehensive framework that can help us identify and understand the processes that govern successful media entertainment. For that purpose, the Media Entertainment Success Cycle is introduced.

The aim of the Media Entertainment Success Cycle (MESC) is to provide an integrative model that combines relevant theories, concepts, and approaches in order to increase conceptual coherence about the processes that dictate entertainment success. Successful entertainment can be defined as a product or content that has achieved positive reception and widespread popularity while generating substantial revenue and profitability for the creators and stakeholders involved. The MESC presents the relations between concepts that shape successful entertainment within two convergent market information processes: industry supply and user demand. In response to user-provided information, the entertainment industry engages in the creation and distribution of content with the intention of providing experiences that satisfy user demands. Simultaneously, entertainment users, through their active participation within existing content supply structures, reshape the contents of these structures to meet changes in their demands. Together, industry and users form a reinforcing market-driven media entertainment cycle in which their mutual preference for popular products stimulates entertainment success. This dynamic supply and demand cycle has been referred to as "media duality," a concept that combines the evolving desires of users with media structures that accommodate those desires (Webster, 2011). The entertainment industry regulates the production of design elements, and the distribution of an entertainment concept, ultimately converging with users in an entertainment experience. The users are influenced by responses to—and effects of—previous products in their preference and motivations for selection of entertainment products. Effective convergence of supply and demand processes culminates in entertainment success.

Success in the entertainment market involves critical acclaim (e.g., recognition, accolades, positive reviews), societal impact (e.g., fans, trendsetting, media attention), evaluative metrics (e.g., views, likes, ratings), and, most importantly, a product's commercial success (e.g., revenue, sales, advertising income). Success predictions are crucial for the entertainment industry as they form the basis for strategic planning and resource allocation, ultimately influencing the reach and impact of entertainment content. The MESC can aid researchers by identifying relevant but underexplored interrelated concepts that can contribute to our understanding of the appeal and effectiveness of media entertainment. The MESC is explained in three steps, starting with the industry supply process outlined in Figure 6.1, followed by the user demand process in Figure 6.2, and the convergence of both processes in Figure 6.3. The complete model, including the culmination in success is displayed in Figure 6.4. In the figures, arrows indicate operational constructs within the supply or demand process and rectangles indicate constructs that are influenced by these processes.

The entertainment industry supply process

The entertainment supply process, displayed in Figure 6.1, emerges from the entertainment industry. The entertainment industry incorporates all sub-industries devoted to entertainment (e.g., film industry, game industry, music industry) and all professional forms of financing, production, distribution, marketing, and exhibition of entertainment products or services therein. The supply process involves several overlapping stages (production, distribution, and marketing) with the aim of ensuring the efficient and effective flow of entertainment products to its users. Production typically starts with the development of an original idea, or the adaptation of an existing concept. Successful entertainment products consistently offer users a sense of familiarity, a perceived positive connection with a product, its concepts, elements, or characters (Bohnenkamp et al., 2015). Familiar elements reduce choice-related uncertainties and offer controlled emotional regulation. One example of this is a viewer repeatedly watching episodes of a familiar sitcom (Russell & Levy, 2012). The appeal of familiar products is also evident in users' preference for music and songs that sound familiar (Ward et al., 2014), as well as higher cinema attendance for movie sequels when compared to non-sequels (Basuroy & Chatterjee, 2008). In order to accommodate users' apparent need for familiar content, many successful entertainment products are extensions of existing brands (e.g., adaptations, covers, sequels, spin-offs, remakes), often distributed across multiple platforms or media channels. A collection of related entertainment products from the same brand, or intellectual property, is called a franchise. For entertainment companies, the benefits of franchises as a successful risk mitigation strategy are beyond question (Chalaby, 2018). The success of any new addition to a franchise relies on the brand's image and awareness, as each extension influences the overall brand perception it also affects the success of future franchise extensions. Although franchise extensions reduce financial risk due to predictable revenue forecasts, financial performance generally diminishes as instalments progress (Filson & Havlicek, 2018). Solely focusing on the exploitation of familiar concepts will inhibit the ability to discover creative and innovative opportunities. Although familiar things tend to be more enjoyable, new things tend to be more interesting (Silvia, 2006). Familiar stimuli can become predictable and boring, whereas novel stimuli that are not fully understood elicit interest, which engages the action tendency of exploring new and exciting experiences (Tan, 2008). Thus, the challenge of successful entertainment production is one of balancing exploitation and exploration: Products need to be appropriately similar in order to feel familiar, yet sufficiently innovative to be distinctive and interesting.

Besides production, distribution is a crucial aspect of the entertainment industry. Distribution is the business of selecting the most suitable method and channel for making a product available to an audience. The objective of successful distribution is to ensure widespread product consumption among diverse (sub)cultures and media. To achieve this, distributors traditionally use the strategic approach known as "windowing," releasing entertainment products over various channels and platforms in succession, aiming to maximise exposure and revenue generation. By

Figure 6.1. The Industry Supply Process in the Media Entertainment Success Cycle

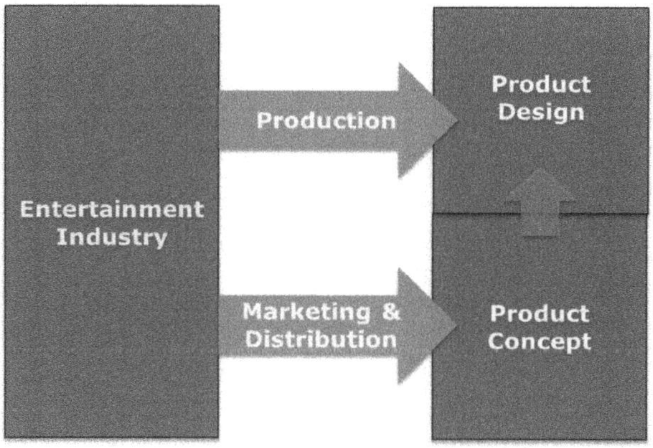

vertically integrating operations, entertainment companies gain more control over the entire value chain, from the creation of content to its delivery to audiences, leading to increased efficiency and profitability through the removal of intermediaries (Hennig-Thurau & Houston, 2019). This disintermediation has also allowed content creators to directly connect with their audience—for instance, through YouTube, TikTok, or Spotify—providing users access to a diverse range of content that aligns with their unique preferences. The "long tail" describes the distribution pattern in which demand for a large number of niche products collectively exceeds the demand for a smaller number of highly popular products. However, niche products face challenges in gaining widespread recognition and profitability, whereas popular products (i.e., blockbusters) have broader appeal and higher likelihood of success, causing the entertainment industry to focus production and distribution efforts on a few concepts with strong mainstream appeal (Elberse, 2013).

For successful distribution, it is essential to align it with a complementing marketing strategy. Marketing involves effectively managing the flow of information about a product. Successful entertainment marketing stimulates buzz, a strong awareness and anticipation among consumers, which boosts success (Thomas, 2004). After release, entertainment products greatly benefit from the perception of success, as consumers tend to rely on rankings when forming quality judgements. Although successful entertainment products are not necessarily better, people tend to perceive them as such, causing initial success to cascade into future success (Elberse & Eliashberg, 2003). Early winners, even by a narrow margin, show huge revenue differences in the long run compared to those in second place (Salganik & Watts, 2008). This cumulative advantage of being perceived as successful is effectively manipulated by film studios by systematically providing news outlets with inflated opening-weekend revenue estimates for their movies (Malhotra & Helmer, 2012). Although heavy users of entertainment select both popular and niche products, this group forms a minority of consumers. The majority of consumers consists of light users, who predominantly select popular products (Elberse, 2013). Thus, successful marketing entails fostering strong awareness and anticipation of popular entertainment concepts, and projecting an image of success.

Prior to watching, playing, or otherwise engaging with entertainment, a product is a "concept," a collection of indicators that allow users to classify and assess what sort of content and design can be expected. Because entertainment products are considered experience goods, they require direct experience to properly evaluate their quality. The cue utilisation theory illustrates that consumers of experience goods rely on specific cues as indicators of product quality, aiding their decision-making process amidst the vast array of available products (Rao & Monroe, 1988). The medium, the genre, and the brand of an entertainment product are three relevant cues, or conceptual indicators, that can create thoughts, emotions, and imagery in the mind of potential users.

First, successfully reaching audiences and conveying information is influenced by the choice of a distribution medium for a product. Television was the dominant medium for the distribution of entertainment 25 years ago, and "the mainstream of the common symbolic environment into which our children are born and in which we all live out our lives" (Gerbner, 1998, p. 177). Currently, the internet is the prime medium for entertainment distribution (e.g., 3.7 million YouTube videos are uploaded every day). Second, genres convey information about the underlying features of a product, guiding users' expectations, and allowing them to make quick decisions. Analyses of Hollywood films released over the last two decades indicates that action and adventure are currently the most-produced film genres (Francemone et al., 2023). Similar to genres, entertainment brands (e.g., Marvel, Pokémon, James Bond) can evoke cognitive associations with a product. Unlike genres, brands are managed professionally. For entertainment products to achieve financial success, heavy investment in a few major brands that have a strong appeal to mass audiences is recommended (Elberse, 2013). Concept indicators like medium, genre, and brand not only influence success by aiding users' classification and potential selection of an entertainment product, they also guide production design choices.

The design of an entertainment product emerges from the production stage and consists of multiple interconnected structural and technological elements that form the language of entertainment. Successful design is highly dependent on the conceptual characteristics of the medium, genre, and brand (e.g., cinematography, user interface, music arrangement, gameplay). The most common design element across entertainment products is some form of narrative structure. In film, narratives typically involve three acts: the setup, the confrontation, and the resolution, with critical plot points placed at the intersections of these acts. Hollywood movies often involve several structured steps in a journey, in which the protagonist who goes on an adventure is met with fierce opposition, overcomes a critical ordeal, emerges victorious is the climax, and returns as a hero (Vogler, 2007). Viewers can appreciate the predictability of narratives that provide structured suspense, arousal, or anguish because of the anticipated emotional redemption that follows the hero's triumph and the antagonist's demise. Indeed, an extensive analysis of screenplays shows that narratives where the protagonist experiences a fall followed by a rise, have more worldwide cinematic success than other narrative arcs, regardless of genre or production budget (Del Vecchio et al., 2021). Its appealing predictability notwithstanding, narratives are but one of the vast amount of design elements that give entertainment products structure and content. Any comprehensive examination of how these different design elements contribute to success is far beyond the scope of the current chapter.

The user demand process

The entertainment demand process, depicted in Figure 6.2, originates from the user and their lifetime of entertainment experiences. Over the last 25 years, research at ASCoR has greatly expanded our knowledge of individual differences in the use of entertainment, and susceptibility to its effects (e.g., Valkenburg & Peter, 2013). These contributions have been pivotal in the conceptualisation of the user demand process of the framework. Users are the consumers, the audience, the viewers, players, or listeners of entertainment products. Individual differences in demographics, personality, and cultural congruence influence preference for specific forms of entertainment. First, demographics such as age and gender can have considerable impact on the use of entertainment. For example, compared to adolescent girls, adolescent boys show a stronger preference for television shows that feature disparaging and slapstick humour (van der Wal et al., 2022). Adolescent boys also show a stronger preference for video games that feature violence (Lemmens et al., 2011a), and adolescent boys are much more likely to expose themselves to sexually explicit forms of online entertainment (Peter & Valkenburg, 2008). Second, personality characteristics, such as traits, temperaments, cognitions, values, attitudes, and beliefs predispose individuals to the selection of specific media. For example, aggressive teens show a stronger preference for televised physical aggression than less aggressive teens (van der Wal et al., 2020), and adolescent gamers with psychosocial vulnerabilities, such as loneliness, low social competence, and low self-esteem, are more likely to develop signs of game addiction (Lemmens et al., 2011b). Third, cultural congruence refers to the alignment between a person's choices and the norms, values, and practices of the culture they are part of. Cultural congruence between an entertainment product and its audience is a powerful determinant of entertainment access and preference. Compared to Dutch boys, boys in the United States attach more value to realistic action and violence (Valkenburg & Jansen, 1999). Intra-culturally, cancel culture has emerged as the public withdrawal of support for celebrities who have acted in a way deemed unacceptable or problematic (e.g., Michael Jackson, J. K. Rowling). The cumulative impact of an individual user's demographics, personality, socio-cultural characteristics, and extensive entertainment experience shapes their preference for particular entertainment content.

"Preference" refers to a user's subjective and dynamic favour or inclination towards certain entertainment concepts over others. Entertainment content is produced and distributed at such a rate that the volume of new material far exceeds our ability to consume those products. The enormous gap between near limitless entertainment and limited attention means it is harder for any product to attract significant attention (Webster, 2011). User preference, determined by tracking user behaviour, can be used for personalised recommendations. Personalised recommendations are one of the most effective ways to influence the selection process and solve the problem of information overload. Recommendation systems are algorithmic tools that use previous entertainment selections to identify and recommend content that users may prefer over other content. These automated personalised recommendations have proven to be very effective at guiding selections among users of Netflix by personalising the images in thumbnails that compile the user interface (Eklund, 2022). These personalised information ecosystems expose users to preferred and familiar content, while shielding them from aversive content. Despite fears that these algorithms would lead to filter bubbles with only homogenous and monotonous content, ASCoR

Figure 6.2. The User Demand Process in the Media Entertainment Success Cycle

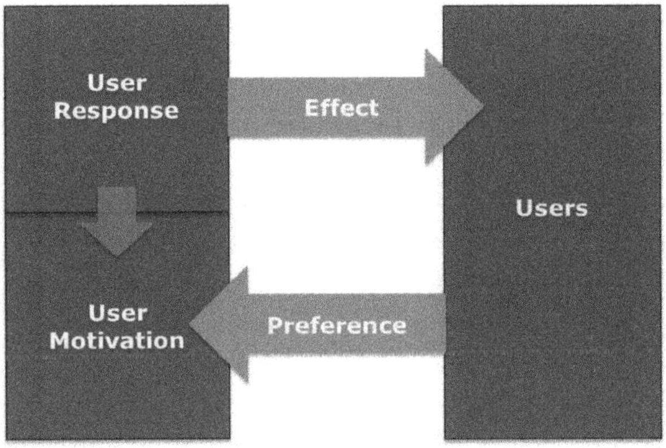

research has shown that most filtering algorithms lead to a diverse set of recommendations (Möller et al., 2020). Not only are filtering algorithms effective at providing preferred yet diverse content, these automated recommendations are trusted as much, or sometimes even more, than those made by human experts (Araujo et al., 2020). While preference facilitates exposure to and ranking of a list of entertainment concepts, the choice for a specific product is often made by the combination of individual motivation and anticipated response.

"Motivation" (see Figure 6.2) describes the tendency or intention to engage in specific goal-directed behaviour, mediating the relation between individual preference and entertainment selection. According to the mood management theory, the motivation for selecting a particular entertainment product is often a function of the expected resulting enjoyment. The selection of hedonic content comes from the motivation to maximise or maintain a good mood, or to diminish or alleviate a bad mood (Zillmann & Bryant, 1985). Alternatively, users may strive for more than mere pleasure and enjoyment, seeking psychological growth through difficult or painful content in eudemonic forms of entertainment that promise mastery, broadening of perspectives, and meaningful self-realisation (Vorderer & Hartmann, 2009). Indeed, search for meaning in life is associated with motivations for selecting eudemonic forms of entertainment, but not hedonic forms of entertainment (Oliver & Raney, 2011). Equilibrium-based theories, such as the uses and gratifications theory, demonstrate that motivations for selecting entertainment should always be regarded in relation to the anticipated emotional, cognitive, and physiological gratifications it will provide. These relations are considered recursive as the evaluation of responses will reinforce or alter audiences' perceptions of gratification-related attributes of particular media content (Rayburn & Palmgreen, 1984). For example, the practice of churning, or definitive disengagement from entertainment products, is affected by unfulfilled starting motivations, such as an unsatisfied need for challenge after repeatedly playing a video game (Daneels et al., 2018).

User "responses" are a combination of emotional, cognitive, and physiological factors that mediate the relation between a media experience and its effects (Valkenburg & Peter, 2013). Enjoyment, the most common response to entertainment, is understood as a positive meta-emotion, meaning that it is a positive emotional response or reaction to a primary emotion. Although enjoyment often emerges from engagement with hedonic content that is light-hearted and fun, leading to primary emotions such as pleasure and amusement, enjoyment can also emerge from content that evokes negative emotions such as fear and sadness. Enjoyment can be distinguished from appreciation through the cognitive involvement required. Enjoyment results from quick, intuitive processing of hedonic content, whereas appreciation comes from eudemonic content that is thought-provoking, moving, or meaningful, causing slower, more controlled appraisals and contemplations (Tamborini, 2011). Eudemonic entertainment experiences are appreciated because they elicit reflection on internalised norms and values and thereby direct attention to meaningful self-realisation through reinterpretation of potentially negative underlying primary emotions (Bartsch & Hartmann, 2017). For example, watching tragedy can break our complacency, make us count our blessings, and realise that things could be a lot worse (Knobloch-Westerwick et al., 2013). More importantly, negative emotions associated with eudemonic content are particularly effective at securing attention, involvement, and ensuring high memorability (Menninghaus et al., 2017).

Media "effects" describe structural changes in a user's knowledge, attitudes, beliefs, physiology, or behaviour resulting from media use. Both hedonic and eudemonic experiences have distinct positive effects on users' psychological well-being. Using hedonic content can ease suffering by encouraging positive prospects, more initiative, and greater resistance to setbacks (Zillmann & Vorderer, 2000). Comparably, long-term use of eudemonic entertainment positively influences psychological resilience and mental health (Reinecke & Kreling, 2022). In social media environments, enjoyment, appreciation or frustration is expressed through evaluative metrics such as ratings, comments, and (dis)likes. These public measures of quality, relevance, and enjoyment effectively influence the attitudes and behaviour of others (Webster, 2011). For example, our research shows that reading positive comments from previous users increases enjoyment of hedonic entertainment content (Möller & Kühne, 2019). Media effects and responses are considered conditional on individual differences (e.g., developmental level, gender), meaning that effects and responses may be stronger among certain groups. The effects of entertainment may also influence further use. Transactional media effects acknowledge the reciprocal interplay between media use and media effects, in which media outcomes influence users' subsequent media selections (Valkenburg & Peter, 2013). For example, time spent playing violent games increases aggressive behavioural tendencies among adolescent boys, which may subsequently stimulate further use of violent games (Lemmens et al., 2011a). In sum, responses to entertainment experiences directly influence motivations for future selections, and indirectly by affecting user preferences.

Converging processes: Selection and experience of entertainment

The convergence of supply-and-demand processes is depicted in Figure 6.3. The supply process concludes in an entertainment experience, which describes the engagement that emerges from the interaction between user and product. The combinations of concept and design elements provide users with engaging experiences that lead to a series of responses. Immersion describes the extent to which an entertainment product is capable of delivering an inclusive, extensive, and vivid illusion of a depicted reality (Slater & Wilbur, 1997). More immersive media experiences generally lead to stronger user responses. Watching a film in a theatre leads to more favourable aesthetic experiences than watching in a home cinema environment (Fröber & Thomaschke, 2021). Immersion is closely related to user engagement experiences (e.g., transportation, involvement, flow) and parasocial experiences (e.g., parasocial interaction, empathy, identification). For example, the experience of presence describes the subjective perception of being in a mediated environment, perceiving the immersive events as really happening (Slater, 2009). Presence can, therefore, best be understood as the experiential counterpart of immersion (Hartmann et al., 2010). ASCoR research on this equivalence has demonstrated that virtual reality (VR) is more immersive than traditional two-dimensional screens because VR gaming induces a stronger sense of presence when compared to a monitor. Furthermore, a stronger sense of presence evoked more intense emotional and physiological responses among players (Lemmens

Figure 6.3. Convergence of Supply and Demand Processes in the Media Entertainment Success Cycle

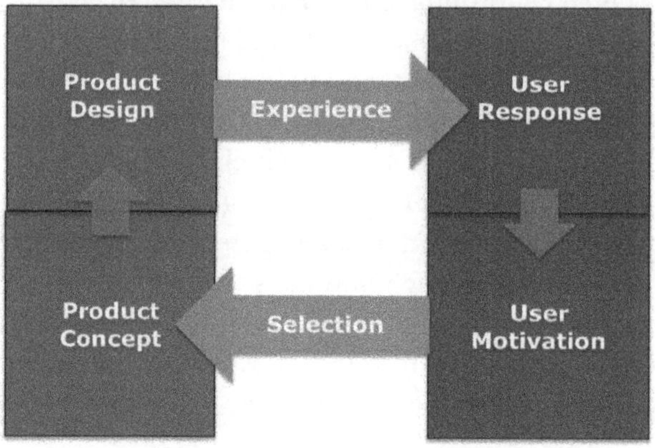

et al., 2022). Similarly, a stronger sense of spatial presence while using an augmented reality app positively affects attitude and behavioural intentions (Smink et al., 2020). In short, industry-controlled immersion leads to more engaging experiences thereby affecting users' responses to entertainment media.

The user demand process concludes with the selection of an entertainment product. Selection dictates the moment, place, and circumstances for users' investment of time, attention, and/or money in entertainment products. Like other media use, entertainment selection becomes predictable and recursive, leading to automaticity after repeated use (LaRose, 2010). Linear entertainment use is bound to patterns of work and leisure, making selection patterns highly predictable. The entertainment industry anticipates these selection patterns, for example, by investing in the production and promotion of content for television's "prime time" and radio's "drive time" (Webster, 2011). Because persistent selection is crucial to the revenue of subscription-based access models (e.g., Disney+, Spotify, Audible), continuous use is encouraged through structural features of the product interface. Netflix's autoplay setting, which requires users to opt-out between episodes of a show, has proven far more effective at stimulating continued engagement than opting-in to start a new episode (Riddle et al., 2018). Viewer data from 12 popular shows on Netflix across 16 countries indicated that users who started the first season of a series, needed to watch an average of 4.25 episodes to keep at least 70% on board for the rest of the season or more (Netflix, 2015). In sum, positive responses resulting from an engaging

Figure 6.4. Successful Supply and Demand Processes in the Media Entertainment Success Cycle

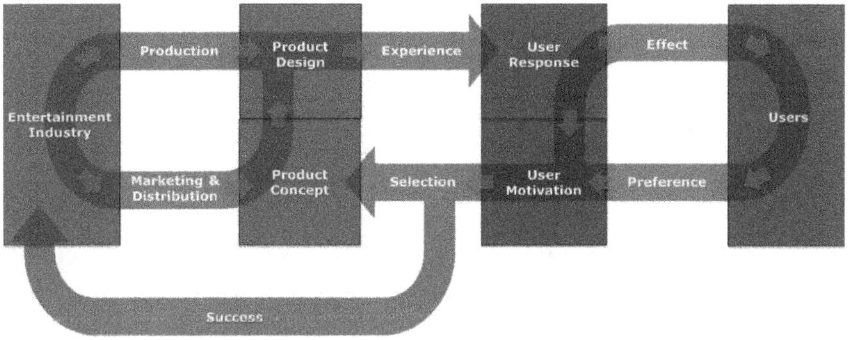

experience motivate repeated selection of the same or similar concepts. The convergence of demand and supply processes thereby contributes to entertainment success, as shown in Figure 6.4.

Successful entertainment

Success in the media entertainment cycle is the positive result from a recursive market-driven framework of mutually reinforcing processes. The dynamics of supply and demand create an equilibrium where popular demand drives creation of popular content, resulting in a seemingly endless loop of similar yet novel products. Popular hedonic concepts (e.g., action & adventure films) lead to more enjoyment, and more financial success, whereas popular eudemonic concepts (e.g., documentary and drama films) lead to more appreciation, critical acclaim, and awards (Francemone et al., 2023). Over the last few decades, the entertainment industry experienced a significant shift in revenue models, largely driven by online advancements and corresponding changes in consumer behaviour. Probably the biggest change occurred in the game industry, where the dominant revenue model shifted from single-game purchases to the monetisation of in-game content. Currently, free mobile games (e.g., *Clash of Clans*, *Candy Crush Saga*, *Fortnite*) with optional in-game purchases, contribute to more than half of the yearly global revenue for the game industry. Specifically, loot boxes have emerged as one of the most effective and profitable forms of game monetisation (Zendle et al., 2020). A loot box is a collective name for different types of packs, chests, or boxes, containing a selection of random items (i.e., loot) that may enhance the gameplay experience. The mechanisms underlying the effectiveness of loot boxes are considered similar to the intermittent reward schedules used in the addictive design of slot machines, making addicted gamers an important source of revenue for the game industry (Lemmens, 2022).

Video revenue models have also changed drastically. Traditional physical distribution has almost completely been replaced by streaming platforms and subscription-based services like Netflix, Prime, and Spotify. Video-sharing platforms like YouTube, Twitch, and TikTok primarily monetise the free distribution of user-generated videos with advertising. Similar to television, revenue mostly comes from brands and companies that advertise with short video ads that appear before, during or after the videos. These platforms may also generate revenue by selling a variety of virtual gifts to their viewers, which they can subsequently gift to their favourite content creators. Creators thereby share in the revenue from these virtual gifts, incentivising them to continue producing more of the same content (Mhalla et al., 2020). Similarly, returns on investments guide the entertainment industry in their plans to extend a franchise or start a new one. For example, decisions on new productions on Netflix are predominantly based on algorithmic feedback provided by users (van Es, 2022). Revenue and feedback thereby determine which content is produced, perpetuating the market-driven cycle of media entertainment success.

Next steps

The MESC indicates that entertainment success research is dominated by studies that focus on product characteristics, whereas ASCoR research primarily examined the individual differences in use of entertainment, user responses, and within-person susceptibility to its effects. Over the past few decades, we have produced a steady stream of highly cited papers that focused on the risks of entertainment, especially among vulnerable populations. Expanding on this, it could be interesting to further examine the ways in which entertainment platforms exploit and manipulate individual preferences, or how certain designs effectively monetise excessive and obsessive use. However, solely fixating on risks may limit our abilities to discover innovative research opportunities. By combining our extensive knowledge of differential susceptibility with the exploration of successful product characteristics and profitable industry practices, we could provide a comprehensive perspective on commercially successful and culturally relevant forms of entertainment. If we keep exploring concepts that present risks and opportunities within emerging forms of successful entertainment, things should remain interesting for at least another 25 years.

References

Araujo, T., Helberger, N., Kruikemeier, S., & de Vreese, C. H. (2020). In AI we trust? Perceptions about automated decision-making by artificial intelligence. *AI & Society, 35*(3), 611–623. https://doi.org/10.1007/s00146-019-00931-w

Asbeek Brusse, E. D., Fransen, M. L., & Smit, E. G. (2015). Educational storylines in entertainment television: Audience reactions toward persuasive strategies in medical dramas. *Journal of Health Communication, 20*(4), 396–405. https://doi.org/10.1080/10810730.2014.965365

Bartsch, A., & Hartmann, T. (2017). The role of cognitive and affective challenge in entertainment experience. *Communication Research, 44*(1), 29–53. https://doi.org/10.1177/0093650214565921

Basuroy, S., & Chatterjee, S. (2008). Fast and frequent: Investigating box office revenues of motion picture sequels. *Journal of Business Research, 61*(7), 798–803. https://doi.org/10.1016/j.jbusres.2007.07.030

Bohnenkamp, B., Knapp, A. K., Hennig-Thurau, T., & Schauerte, R. (2015). When does it make sense to do it again? An empirical investigation of contingency factors of movie remakes. *Journal of Cultural Economics, 39*(1), 15–41. https://doi.org/10.1007/s10824-014-9221-6

Boukes, M., Boomgaarden, H. G., Moorman, M., & de Vreese, C. H. (2015). At odds: Laughing and thinking? The appreciation, processing, and persuasiveness of political satire. *Journal of Communication, 65*(5), 721–744. https://doi.org/10.1111/jcom.12173

Chalaby, J. K. (2018). Hedging against disaster: Risk and mitigation in the media and entertainment industries. *International Journal of Digital Television, 9*(2), 167–184. https://doi.org/10.1386/jdtv.9.2.167_1

Daneels, R., Malliet, S., Koeman, J., & Ribbens, W. (2018). The enjoyment of shooting games: Exploring the role of perceived realism. *Computers in Human Behavior, 86*, 330–336. https://doi.org/10.1016/j.chb.2018.04.053

Del Vecchio, M., Kharlamov, A., Parry, G., & Pogrebna, G. (2021). Improving productivity in Hollywood with data science: Using emotional arcs of movies to drive product and service innovation in entertainment industries. *Journal of the Operational Research Society, 72*(5), 1110–1137. https://doi.org/10.1080/01605682.2019.1705194

Eklund, O. (2022). Custom thumbnails: The changing face of personalisation strategies on Netflix. *Convergence, 28*(3), 737–760. https://doi.org/10.1177/13548565211106452

Elberse, A. (2013). *Blockbusters: Hit-making, risk-taking, and the big business of entertainment.* Henry Holt and Co.

Elberse, A., & Eliashberg, J. (2003). Demand and supply dynamics for sequentially released products in international markets: The case of motion pictures. *Marketing Science, 22*(3), 329–354. https://doi.org/10.1287/mksc.22.3.329.17740

Filson, D., & Havlicek, J. H. (2018). The performance of global film franchises: Installment effects and extension decisions. *Journal of Cultural Economics, 42*(3), 447–467. https://doi.org/10.1007/s10824-018-9316-6

Francemone, C. J., Kryston, K., & Grizzard, M. (2023). Examining genre success, co-occurrence, release, and production of 9,068 films over twenty years. *Poetics, 99*, Article 101794. https://doi.org/10.1016/j.poetic.2023.101794

Fröber, K., & Thomaschke, R. (2021). In the dark cube: Movie theater context enhances the valuation and aesthetic experience of watching films. *Psychology of Aesthetics, Creativity, and the Arts, 15*(3), 528–544. https://doi.org/10.1037/aca0000295

Gerbner, G. (1998). Cultivation analysis: An overview. *Mass Communication and Society, 1*(3–4), 175–194. https://doi.org/10.1080/15205436.1998.9677855

Hartmann, T., Klimmt, C., & Vorderer, P. (2010). Telepresence and media entertainment. In C. C. Bracken & P. Skalski (Eds.), *Immersed in media: Telepresence in everyday life* (pp. 137–157). Routledge. https://research.vu.nl/ws/portalfiles/portal/2764210/Bracken+C007.pdf

Hennig-Thurau, T., & Houston, M. B. (2019). *Entertainment science: Data analytics and practical theory for movies, games, books, and music.* Springer Nature. https://doi.org/10.1007/978-3-319-89292-4

Knobloch-Westerwick, S., Gong, Y., Hagner, H., & Kerbeykian, L. (2013). Tragedy viewers count their blessings: Feeling low on fiction leads to feeling high on life. *Communication Research, 40*(6), 747–766. https://doi.org/10.1177/0093650212437758

LaRose, R. (2010). The problem of media habits. *Communication Theory, 20*(2), 194–222. https://doi.org/10.1111/j.1468-2885.2010.01360.x

Lemmens, J. S. (2022). Play or pay to win: Loot boxes and gaming disorder in *FIFA Ultimate Team. Telematics and Informatics Reports, 8*, Article 100023. https://doi.org/10.1016/j.teler.2022.100023

Lemmens, J. S., Simon, M., & Sumter, S. R. (2022). Fear and loathing in VR: The emotional and physiological effects of immersive games. *Virtual Reality, 26*(1), 223–234. https://doi.org/10.1007/s10055-021-00555-w

Lemmens, J. S., Valkenburg, P. M., & Peter, J. (2011a). The effects of pathological gaming on aggressive behavior. *Journal of Youth and Adolescence, 40*(1), 38–47. https://doi.org/10.1007/s10964-010-9558-x

Lemmens, J. S., Valkenburg, P. M., & Peter, J. (2011b). Psychosocial causes and consequences of pathological gaming. *Computers in Human Behavior, 27*(1), 144–152. https://doi.org/10.1016/j.chb.2010.07.015

Malhotra, N., & Helmer, E. (2012). Inflation in weekend box office estimates. *Applied Economics Letters, 19*(14), 1411–1415. https://doi.org/10.1080/13504851.2011.631883

Menninghaus, W., Wagner, V., Hanich, J., Wassiliwizky, E., Jacobsen, T., & Koelsch, S. (2017). The Distancing-Embracing model of the enjoyment of negative emotions in art reception. *The Behavioral and Brain Sciences, 40*, Article e347. https://doi.org/10.1017/S0140525X17000309

Mhalla, M., Jiang, Y., & Nasiri, A. (2020). Video-sharing apps business models: TikTok case study. *International Journal of Innovation and Technology Management, 17*(7), Article 2050050. https://doi.org/10.1142/S0219877020500509

Möller, A. M., & Kühne, R. (2019). The effects of user comments on hedonic and eudaimonic entertainment experiences when watching online videos. *Communications: The European Journal of Communication Research, 44*(4), 427–446. https://doi.org/10.1515/commun-2018-2015

Möller, J., Trilling, D., Helberger, N., & van Es, B. (2020). Do not blame it on the algorithm: An empirical assessment of multiple recommender systems and their impact on content diversity. In M. Beaufort (Ed.), *Digital media, political polarization and challenges to democracy* (pp. 45–63). Routledge. https://doi.org/10.1080/1369118X.2018.1444076

Netflix. (2015, September 23). *Do you know when you were hooked? Netflix does.* Netflix Newsroom. https://about.netflix.com/en/news/do-you-know-when-you-were-hooked-netflix-does

Oliver, M. B., & Raney, A. A. (2011). Entertainment as pleasurable and meaningful: Identifying hedonic and eudaimonic motivations for entertainment consumption. *Journal of Communication, 61*(5), 984–1004. https://doi.org/10.1111/j.1460-2466.2011.01585.x

Peter, J., & Valkenburg, P. M. (2008). Adolescents' exposure to sexually explicit internet material, sexual uncertainty, and attitudes toward uncommitted sexual exploration: Is there a link? *Communication Research, 35*(5), 579–601. https://doi.org/10.1177/0093650208321754

Rao, A. R., & Monroe, K. B. (1988). The moderating effect of prior knowledge on cue utilization in product evaluations. *Journal of Consumer Research, 15*(2), 253–264. https://doi.org/10.1086/209162

Rayburn, J. D., & Palmgreen, P. (1984). Merging uses and gratifications and expectancy-value theory. *Communication Research, 11*(4), 537–562. https://doi.org/10.1177/009365084011004005

Reinecke, L., & Kreling, R. (2022). The longitudinal influence of hedonic and eudaimonic entertainment preferences on psychological resilience and well-being. *Frontiers in Communication, 7*, Article 991458. https://doi.org/10.3389/fcomm.2022.991458

Riddle, K., Liao, X., & Minich, M. (2022). Media enjoyment: A synthesis. *Journal of Broadcasting & Electronic Media, 66*(4), 592–622. https://doi.org/10.1080/08838151.2022.2137510

Riddle, K., Peebles, A., Davis, C., Xu, F., & Schroeder, E. (2018). The addictive potential of television binge watching: Comparing intentional and unintentional binges. *Psychology of Popular Media Culture, 7*(4), 589–604. https://doi.org/10.1037/ppm0000167

Russell, C. A., & Levy, S. J. (2012). The temporal and focal dynamics of volitional reconsumption: A phenomenological investigation of repeated hedonic experiences. *Journal of Consumer Research, 39*(2), 341–359. https://doi.org/10.1086/662996

Salganik, M. J., & Watts, D. J. (2008). Leading the herd astray: An experimental study of self-fulfilling prophecies in an artificial cultural market. *Social Psychology Quarterly, 71*(4), 338–355. https://doi.org/10.1177/019027250807100404

Silvia, P. J. (2006). *Exploring the psychology of interest.* Oxford University Press. https://doi.org/10.1093/acprof:oso/9780195158557.001.0001

Slater, M. (2009). Place illusion and plausibility can lead to realistic behaviour in immersive virtual environments. *Philosophical Transactions of the Royal Society of London, 364*(1535), 3549–3557. https://doi.org/10.1098/rstb.2009.0138

Slater, M., & Wilbur, S. (1997). A framework for immersive virtual environments (FIVE): Speculations on the role of presence in virtual environments. *Presence: Teleoperators & Virtual Environments, 6*(6), 603–616. https://doi.org/10.1162/pres.1997.6.6.603

Smink, A. R., van Reijmersdal, E. A., van Noort, G., & Neijens, P. C. (2020). Shopping in augmented reality: The effects of spatial presence, personalization and intrusiveness on app and brand responses. *Journal of Business Research, 118*, 474–485. https://doi.org/10.1016/j.jbusres.2020.07.018

Tamborini, R., Grizzard, M., Bowman, N. D., Reinecke, L., Lewis, R. J., & Eden, A. (2011). Media enjoyment as need satisfaction: The contribution of hedonic and nonhedonic needs. *Journal of Communication, 61*(6), 1025–1042. https://doi.org/10.1111/j.1460-2466.2011.01593.x

Tan, E. S. H. (2008). Entertainment is emotion: The functional architecture of the entertainment experience. *Media Psychology, 11*(1), 28–51. https://doi.org/10.1080/15213260701853161

Thomas Jr, G. M. (2004). Building the buzz in the hive mind. *Journal of Consumer Behaviour: An International Research Review, 4*(1), 64–72. https://doi.org/10.1002/cb.158

Valkenburg, P. M., & Janssen, S. C. (1999). What do children value in entertainment programs? A cross-cultural investigation. *Journal of Communication, 49*(2), 3–21. https://doi.org/10.1111/j.1460-2466.1999.tb02790.x

Valkenburg, P. M., & Peter, J. (2013). The differential susceptibility to media effects model. *Journal of Communication, 63*(2), 221–243. https://doi.org/10.1111/jcom.12024

van Berlo, Z. M. C., van Reijmersdal, E. A., & Eisend, M. (2021). The gamification of branded content: A meta-analysis of advergame effects. *Journal of Advertising, 50*(2), 179–196. https://doi.org/10.1080/00913367.2020.1858462

van der Wal, A., Fikkers, K. M., & Valkenburg, P. M. (2020). What's in it for them? Teens' differential preferences for types and contexts of televised aggression. *Communication Research, 47*(8), 1206–1227. https://doi.org/10.1177/0093650219832231

van der Wal, A., Pouwels, J. L., Piotrowski, J. T., & Valkenburg, P. M. (2022). Just a joke? Adolescents' preferences for humor in media entertainment and real-life aggression. *Media Psychology, 25*(6), 797–813. https://doi.org/10.1080/15213269.2022.2080710

van Es, K. (2022). Netflix & big data: The strategic ambivalence of an entertainment company. *Television & New Media*. Advance online publication. https://doi.org/10.1177/15274764221125745

van Leeuwen, L., van den Putte, B., Renes, R. J., & Leeuwis, C. (2017). Do narrative engagement and recipients' thoughts explain the impact of an entertainment-education narrative on discouraging binge drinking? *Media Psychology, 20*(2), 194–220. https://doi.org/10.1080/15213269.2016.1142379

Vermeer, S., Trilling, D., Kruikemeier, S., & de Vreese, C. (2020). Online news user journeys: The role of social media, news websites, and topics. *Digital Journalism, 8*(9), 1114–1141. https://doi.org/10.1080/21670811.2020.1767509

Vogler, C. (2007). *The writer's journey: Mythic structure for writers* (3rd ed.). Michael Wiese Productions.

Vorderer, P., & Hartmann, T. (2009). Entertainment and enjoyment as media effects. In J. Bryant & M. B. Oliver (Eds.), *Media effects* (3rd ed., pp. 532–550). Routledge. https://research.vu.nl/ws/portalfiles/portal/2497888/2008+Vorderer+Hartmann+-+Entertainment+Media+Effects.pdf

Ward, M. K., Goodman, J. K., & Irwin, J. R. (2014). The same old song: The power of familiarity in music choice. *Marketing Letters, 25*(1), 1–11. https://doi.org/10.1007/s11002-013-9238-1

Webster, J. G. (2011). The duality of media: A structurational theory of public attention. *Communication Theory, 21*(1), 43–66. https://doi.org/10.1111/j.1468-2885.2010.01375.x

Zendle, D., Meyer, R., & Ballou, N. (2020). The changing face of desktop video game monetisation: An exploration of exposure to loot boxes, pay to win, and cosmetic microtransactions in the most-played Steam games of 2010–2019. *PloS ONE, 15*(5), Article e0232780. https://doi.org/10.1371/journal.pone.0232780

Zillmann, D., & Bryant, J. (1985). Affect, mood, and emotion as determinants of selective exposure. In D. Zillmann & J. Bryant (Eds.), *Selective exposure to communication* (pp. 157–190). Lawrence Erlbaum Associates. https://doi.org/10.4324/9780203056721

Zillmann, D., & Vorderer, P. (2000). *Media entertainment: The psychology of its appeal.* Routledge. https://doi.org/10.4324/9781410604811

Acknowledgements

Several people have helped in the preparation of this chapter. First, I would like to thank Patti M. Valkenburg and Jochen Peter for their inspiring DSMM. I would also like to thank Sindy Sumter and the editors for their valuable comments and suggestions. Special thanks go to Susanne Baumgartner and Tilo Hartmann for their insightful and innovative suggestions. Thanks to Mieke for her patience and Pieter for the graphics.

About the author

Jeroen Lemmens is an Assistant Professor at ASCoR specialising in game addiction and VR gaming. His research is aimed at understanding the role of media entertainment by focusing on the risks and opportunities provided by old and emerging forms of digital entertainment. E-mail: j.s.lemmens@uva.nl

7. Organisations, Media, and Society

Toni G. L. A. van der Meer, Christian Burgers, Sandra H. J. Jacobs, Pytrik Schafraad, and Mark Boukes

Abstract

The dynamic interplay between organisations, media, and society is central to this chapter and highlights ASCoR's Corporate Communication group's issue-centred and society-focused approach to explore communication between organisations and their environment. Studying this interplay allows questions to be answered regarding organisations' role in our mediated society, how organisations shape and are shaped by public and media debates, and how societal issues have become inherently intertwined with organisational practices. To provide a broad overview of the studies conducted by ASCoR researchers on this specific topic, this chapter outlines three main topics: (1) organisations and news media, (2) mediatisation of organisations, and (3) organisational legitimacy.

Keywords: corporate communication, issue communication, news media, mediatisation of organisations, organisational legitimacy

Introduction

What role do organisations play in our mediated society? How do organisations shape and how are they shaped by public and media debates? These are key themes central to the research within ASCoR's Corporate Communication group. Through processes of mediatisation, media have become institutionalised in all layers of society, including the operations of all types of organisations (Ihlen & Pallas, 2014), ranging from governmental organisations to non-governmental organisations (NGOs) and for-profit corporations. This development has largely shaped how organisations communicate with their stakeholders and take up their roles in society.

---

van der Meer, T. G. L. A., Burgers, C., Jacobs, S. H. J., Schafraad, P., & Boukes, M. (2024). Organisations, media, and society. In T. Araujo & P. Neijens (Eds.), *Communication research into the digital society: Fundamental insights from the Amsterdam School of Communication Research* (pp. 121-138). Amsterdam University Press. https://doi.org/10.5117/9789048560592_ch07

Over the last decades, the operating conditions for organisations have been decisively shaped by preferences and practices of the media. In this chapter, the term "media" primarily refers to news media, rather than social media, in order to grasp how organisations have, through processes of mediatisation, become more related to and intertwined with their environment and therewith the issues that play in society more broadly. Organisations today cannot ignore or avoid media logic to reach their stakeholders and adapt to their environment. With these changes, societal visibility of organisations and their activities have increased. Adherence to such media and public scrutiny is, for example, expressed through growing corporate communication, media relations, and public relations departments in organisations (Ihlen & Pallas, 2014).

Since organisations have become inherently intertwined with their mediatised surroundings, the role of communication has changed for organisations. Organisations now have to closely monitor the media, political, and public agendas in relation to their operations, and participate in issue arenas on societal topics relevant to them (van der Meer & Jonkman, 2021). The communicative interplay between organisations and their environment helps to shape, maintain, and renew their social licence to operate (SLO), which is broadly understood as a form of soft regulation of organisations and whether their actions are deemed acceptable in the eyes of relevant stakeholders in society. Organisational legitimacy has become a crucial element that is believed to be largely constructed by media. The important role that organisations play in society is therewith primarily negotiated and defined through mediated communication. Accordingly, the study of the dynamic interplay between organisations, media, and society is central to the research done within ASCoR's Corporate Communication group. We gather all communication by or about organisations under the umbrella term "corporate communication." Here the term "organisation" is understood broadly to contain for-profit corporations, governmental actors, nonprofit organisations or activist groups, and (news) media organisations. We rely on an issue-centred approach to empirically study the communicative interplay between organisations, media, and society, which allows us to capture organisations' role within the mediated society. In line with the perspective of issues arenas (Luoma-aho & Vos, 2010), we understand this issue-centred approach as issues being central to organisational communication. Just like other actors, organisations can have a stake in an issue, which connects them to other involved actors. Media provide the primary platforms where organisations meet and engage with their environment to negotiate these issues, ranging from crises restricted to specific organisations (e.g., legitimacy concerns, product recall, rumours) to largely societal issues (e.g., sustainability, diversity, polarisation, misinformation).

To understand the interplay between organisations, media, and society, research conducted in ASCoR's Corporate Communication group is addressed in this chapter. It is important to note that ASCoR primarily relies on an empirical-positivistic approach to study this interplay, rather than alternative perspectives of studying corporate communication (e.g., critical studies) that have also provided the substantial contribution to literature that our research has relied upon. As a first step, it is important to understand the role that news media plays in organisations' visibility. To elaborate on organisational visibility, we address research that has been conducted on the way that the news media have reported about organisations and, subsequently, what effects this media attention may have for these organisations. Thereby, we follow Verhoeven's (2016) literature review which identified several variables central in the study of corporate communication: message credibility, perceived trustworthiness, reputation, consumer confidence, and stock prices.

Secondly, we discuss research on the mediatisation of organisations to understand, on a deeper level, how media have become institutionalised in organisational processes and operations. "Mediatisation" was originally coined as a term to capture the impact of media on politics. It can be understood as a long-term, multi-phase process that ranges from media as an important source of information for political actors to more profound changes like "political actors being governed by media logic" (Strömbäck, 2008). The scope of the mediatisation concept soon expanded to organisations to capture "the significance of the media" and its intervention in organisational life (Fredriksson & Pallas, 2021, p. 168). Over the years, mediatisation has proven to be an applicable perspective to understand the institutionalisation of media in organisations' communicative processes and activities (e.g., Ihlen & Pallas, 2014). In the organisational context, the phenomenon of mediatisation helps us understand how media decisively shape the communicative environment of organisations and how other institutions such as organisations and their members adapt to the logic of the media.

Finally, to understand how the communicative interplay with media and society shapes organisations, we address organisational legitimacy. Legitimacy is a crucial, yet vulnerable resource for organisations. As legitimacy refers to a stakeholders' perceived congruence of an organisation with societal norms, focusing on the role of the media in the reflection and co-construction of organisational legitimacy is paramount (Dekker et al., 2021; van der Meer & Jonkman, 2021). Recent transformations in the media landscape have made the case for a communication perspective on legitimacy even stronger, as social evaluations of organisations are currently more visible, fragmented, and have the potential to amplify quickly (Etter et al., 2019; van der Meer & Jonkman, 2021). We understand legitimacy as a fluid concept: Legitimacy is not given by default, is subject to repeated change, and needs to be constantly renewed. Not only does organisational legitimacy needs to be adapted to current changes in societal norms, it also needs to consider conflict management, critical risk, and the crisis issues constantly confront organisations' place within society. Hence, the communicative construction of organisational legitimacy has become a central topic for both corporate communication practitioners and scholars (Marschlich, 2022).

In the next section, we address research conducted within ASCoR's Corporate Communication group by addressing the three topics above by first discussing methods and findings, followed by contributions to theory and practice, and a future research agenda.

Empirical findings

As a start to this chapter, we outline the methodological approaches applied in ASCoR's Corporate Communication group. Alongside the discussion of the variety of methods, the unique insights regarding the interplay between organisations, media, and society that have been brought forward by the group are summarised. To structure the methods and findings section, the three topics introduced above are discussed separately.

News media and organisations

Various studies conducted by ASCoR's Corporate Communication group have investigated the interdependencies between organisations and news media on the micro level (i.e., the individual level) (Jacobs & Wonneberger, 2017) as well as the meso (i.e., organisational level) and macro levels (i.e., societal level) (e.g., Kroon & van der Meer, 2021; Schafraad et al., 2016; Strauß & Vliegenthart, 2017). In a study that combined survey with content analysis, Jacobs and Wonneberger (2017) found that media orientation of communication professionals of organisations—i.e., how important they believe it is for their organisation to attract media attention and how this influences strategic decision-making—depends more on their organisation's perceived visibility in the media rather than its actual media visibility.

Several content analyses of news media coverage of organisations conducted by ASCoR's Corporate Communication group focus on actual visibility of organisations (or actors) in news media and their predictability and show that the sheer number of news factors such as conflict and eliteness present predict prominence and size of the news report (Boukes et al., 2022). In a similar way, Jonkman et al. (2020) found that corporate characteristics, such as company size and ownership structure, predict organisations' visibility in news reports.

Other studies focused on the first-level agenda-building role of organisations' press releases; these studies found a significant impact of press releases on the topics of business news in both broadsheet and tabloid news outlets (Kroon & van der Meer, 2021), where press releases containing multiple news factors draw most news media attention (Schafraad et al., 2016). Online news outlets, which rely on the 24/7 news cycle and need to be "first with the news," especially depend on these press releases, whereas the press releases' impact on traditional media, such as newspapers and television news, is weaker (Boumans et al., 2018).

After passing the gatekeepers of the news agenda (i.e., visibility), organisations have less influence on the content and tone of news coverage. Studies based on manual as well as automated content analysis of both organisations' press releases and news media content demonstrate that, in the news-gathering phase, news media rely much less on such forms of information subsidies provided by organisations through press releases: The more news factors present in press releases, the more original work journalists put into the news report (Schafraad & van Zoonen, 2020) and affective attribute salience (i.e., sentiment) in press releases only have a small effect on the news agenda (Kroon & van der Meer, 2021).

Moreover, using a computational framing approach combined with time series analysis, Strauß and Vliegenthart (2017) found that, in the context of the banking crisis, there is a reversed agenda-building effect with news media frames influencing corporate communication frames rather than news media following corporate communication. In another study, Strauß et al. (2016) found that neither positive nor negative emotions in newspaper coverage consistently affect the values of stocks listed on the Amsterdam Exchange index (AEX), but, instead, that media coverage itself was likelier to be influenced by the fluctuations on the stock market.

Mediatisation of organisations

The research programme of ASCoR's Corporate Communication group has shown the theoretical and empirical relevance of the concept of mediatisation (i.e., the institutionalisation of media with society over time) to study the relationship between the organisation and their environment. The mediatisation of organisations is studied on two distinctive levels, namely: (1) how organisations are monitored and covered by the media and (2) how organisations and their member adapt to media logic. First, the way organisations are evaluated by society is increasingly constructed in the media and how organisations adapt to media is crucial for them to reach their stakeholders. The media's monitoring and portrayal of organisations has changed over the last decades. Through media logic, organisations are increasingly scrutinised by news media and their coverage increasingly follows a media logic (for example, a focus on news values such as negativity, conflicts, sensationalism) to gain a larger audience. On this level, content analyses provide a particular valuable instrument to understand organisational visibility and how organisations are framed by news media. Research over time especially allows for exploring how media coverage of organisations has changed against the backdrop of long-term mediatisation processes.

Empirical research has shown how the media visibility of organisations changed and they became more media oriented. First, as addressed above, the communicative agendas of organisations and the news are found to be intertwined in an intimate relation of reciprocal influence (Kroon & van der Meer, 2021) and the media agenda can shape organisations' evaluative features such as their stock market ratings (van der Meer & Vliegenthart, 2018). More related to the processes of mediatisation over time, the way that media cover organisation-related issues is found to change with time regarding certain issues. For example, in the case of aviation accidents, media tend to follow their own logic rather than real-world data, since relative media attention for these negative events went up over time (and increasingly contained mediatisation facets related to elements such as conflict framing, sensationalism of news, and human interest framing) while the total number of aviation incidents declined (van der Meer et al., 2019; van der Meer et al., 2022). This mediatised coverage can have effects on the organisational level (i.e., decreasing air travel) and societal level (more deadly road accidents as result of irrational risk avoiding travel behaviour). Moreover, also under specific organisational circumstances, media are found to play a central and decisive role. For example, the majority of external organisational crisis situations can be understood as mediatised situations due to the central and leading role of news media in the communicative interplay between organisations and their environment at these critical times (van der Meer, 2016; van der Meer et al., 2014). News media's central role during crises becomes evident from being the main source for, and influencer of, public crisis framing (van der Meer, 2018) and having the potential to have a soothing effect on public panic and speculation (van der Meer & Verhoeven, 2013) or ease the crisis atmosphere (Gerken & van der Meer, 2019).

Second, next to these macro- and meso-level changes of mediatisation, Nölleke et al. (2021) discern micro-level (individual) adaptation. Here, qualitative (interviews) and quantitative (survey) research among communication professionals provide insights into their organisations' and their own media orientation (Jacobs & Wonneberger, 2019; Wonneberger & Jacobs, 2016). Inspired by Kohring and colleagues' (2013) conceptualisation of mental mediatisation, media orientation can be considered as mediatisation manifested at the individual level within organisations. In other words, the extent to which communication professionals are oriented towards media relations and logic can be an indication of the institutionalisation of media with organisations. This concept has been used to describe media-related attitudes of organisational members concerning the media's impact on and its importance for their organisation (Jacobs & Wonneberger, 2019, p. 921; Wonneberger & Jacobs, 2016). A qualitative interview study resulted in a conceptualisation consisting of three interacting dimensions: the perceived function of media coverage for the organisation (attention seeking), the perceived relevance of specific medium types for the organisation and its decision-making processes (strategic impact), and the evaluation of media coverage about the organisation and the media environment (media hostility) (Jacobs & Wonneberger, 2019, p. 927). It can be used to assess changes over time in organisational members' media-related attitudes, which are expected to play a role in the formation of the organisation's strategic communication policies. Furthermore, in today's mediatised society, we see how communication professionals can feel particularly high pressure from news media (van der Meer et al., 2017).

Organisational legitimacy

When it comes to social evaluations of organisations, reputation has traditionally been the core target of communication efforts by practitioners and studied as a key outcome variable in academic research (Jacobs &

Liebrecht, 2023). While reputation is competitive and performance-oriented, legitimacy stresses perceived congruence with societal norms or with other, similar types of organisations (Deephouse & Suchman, 2008). Corporate communication's recent shift towards a more societal and issue-based orientation has likewise shifted the attention to organisations' legitimacy and its interrelations with strategic communication and media (van der Meer & Jonkman, 2021). A lack of legitimacy is troublesome for an organisation's position in society and negatively affects its chances for long-term survival (Waeraas, 2020).

Media coverage plays a constitutive role in organisational legitimacy by presenting legitimacy perceptions held by stakeholder groups and by simultaneously co-constructing these perceptions by "providing the battleground" to discuss organisations' congruence with societal norms (Marschlich & Ingenhoff, 2022). A recent empirical account of media legitimacy can be found in the work of Marschlich and Ingenhoff (2022), who conducted a manual quantitative content analysis of newspaper coverage in the United Arab Emirates to assess news media framing of corporate diplomacy—i.e., organisations' communicative engage aimed at influencing sociopolitical trends in order to gain legitimacy—by multinational companies. Their analysis revealed that two out of three identified media frames focused on legitimacy elements related to the distribution of organisational benefits. Moral media legitimacy refers to media coverage that portrays the benefits of corporate diplomacy as distributed among communities or society at large or as congruent with societal norms and values, whereas "pragmatic media legitimacy occurs when news coverage emphasises the benefits of corporate diplomacy-related activities for individual actors and even groups within society at large" (Marschlich & Ingenhoff, 2022, p. 5). In sum, Marschlich and Ingenhoff (2022) lay out an empirical, content-analytical approach to media legitimacy which can be adopted in similar types of studies.

Besides this empirical account, a substantial amount of communication-scientific work on media legitimacy has a normative and/or conceptual orientation (e.g., Waeraas, 2020). This emphasises the importance of conducting further empirical work on this topic (see also Marschlich, 2022, p. 14). For that reason, much empirical work in the field of media legitimacy is currently ongoing. Current empirical projects in ASCoR's Corporate Communication group aim to analyse mediated organisational legitimacy constructions from various methodological angles. For instance, questions we aim to answer relate to the different types of legitimacy frames that are used in news media to cover public sector organisations, how news organisations retain their legitimacy in the light of societal challenges in the digital age, and how issue communication strategies affect corporate reputation and/or legitimacy.

Contributions to theory and practice

Based on the summary of the findings above, below we outline ASCoR's Corporate Communication group's contributions to the field. Again, we rely on the three topics to organise the contributions.

Media and organisations

The studies conducted within ASCoR's Corporate Communication group on the coverage of corporations in the news have enriched various theoretical frameworks, such as agenda-setting theory (e.g., Kroon & van der Meer, 2021; Strauß & Vliegenthart, 2017) and news values theory (e.g., Boukes et al., 2022; Schafraad et al., 2016). Furthermore, other studies in this domain have demonstrated the *direct* impact and relevance of this media coverage for the involved corporations. For instance, several studies demonstrated how both newspaper coverage (van der Meer & Vliegenthart, 2018) and press agency tweets (Strauß et al., 2018), impacted corporation's stock prices. In particular, media that are constantly updated (e.g., social media) can be particularly impactful in this regard. This demonstrates that new media may be more impactful for the fast-paced work of traders than traditional newspaper coverage, which by definition is always delayed. Future research on this topic can thus further investigate how media coverage of organisations in traditional (mass) media may have differential societal effects compared to media coverage in social media.

While many studies within ASCoR's Corporate Communication group on the connection between media and organisations focus on societal-level effects, it is also important to zoom in on individual effects (e.g., by combining content-analytic and survey data). For instance, Jonkman et al. (2020) demonstrated how particularly negatively valanced news can damage the reputation of organisations, but only for some news consumers. After all, they found a buffering effect through which people with initially positive attitudes about a corporation were less influenced by negative news. Altogether, this underlines the need for future research to also pay attention to characteristics of individual consumers and social groups and when assessing the impact of media coverage of organisations.

Mediatisation

The work on mediatisation by ASCoR's Corporate Communication group has provided a theoretical fundament for understanding how organisations' operating conditions, but also their public legitimacy, are largely constructed within the media debate. Through processes of mediatisation, organisational issues are shaped and given meaning through communication on platforms provided by the media. The centrality of media for organisations' interaction with their surroundings highlight how organisations cannot avoid aligning their communication according to the rules of the media. Organisations have to embrace the media logic and are often forced to adopt to changes in the media landscape. For example, new developments like the rise of (fear for) misinformation (van der Meer & Jin, 2020), increased polarisation of the public debate (van der Meer & Jonkman, 2021), and the increasing powerful role of Big Tech companies in our information environment (Schwinges et al., 2022) are challenges of the media landscape that cannot be ignored by organisations. Overall, the research on mediatisation highlight how we cannot understand organisations' functions within society without taking the role of the media into account.

Legitimacy

In their reflection on organisations' legitimacy, van der Meer and Jonkman (2021) argue that corporations should also reflect on their SLO, which can be seen as the ways in which an organisation's social environment perceives and evaluates how the organisation deals with sociopolitical issues. Van der Meer and Jonkman (2021) state that both mediatisation and politicisation (i.e., the fact that organisations are asked more often to take an explicit position on sociopolitical issues) of corporations' environments push organisations towards the renewal of their SLO. Their work recognises the key role that legitimacy considerations do and should play in corporate communication efforts. Analyses of the content and effects of media legitimacy strategies can inform legitimacy-building and strategic communication efforts by organisations (Marschlich & Ingenhoff, 2022). The research projects currently being carried out will also contribute to a growing set of insights related to media legitimacy as an empirical phenomenon. The often-used definition of legitimacy as a multidimensional construct evaluated by diverse audiences (Marschlich & Ingenhoff, 2022) provides a solid basis for further empirical explorations.

Next steps

The large body of varied work conducted within ASCoR's Corporate Communication group on the relation between organisations, media, and society has revealed how media play an important part in the ways in which perceptions and behaviours of organisations are shaped. News media can place organisations (and topics relevant to them) on the public, media, or political agendas, and, in these ways, affect the public's view on these organisations. The group's research on mediatisation shows how media are thus crucial in shaping perceptions by and of organisations. Through mediatisation and politicisation, organisations are required to participate more often in issue arenas on societal topics, even when organisations themselves would not necessarily want to engage in these issue arenas. In this way, we see a societal focus that requires that organisations do not solely focus on shareholder value, but—instead—also explicitly indicate how they operate for the benefit in society. In case organisations fail to do so, their legitimacy and their SLO may be put up for discussion in (social) media.

Future ASCoR research on these themes will focus on a variety of topics. First and foremost, the current lines of research will be continued and future work will further uncover when, how, and for whom media coverage of issue arenas impact perceptions of organisations. In addition, future work will further specify how mediatisation processes impact the operation of organisations and indicate how legitimacy framing is crucial for organisations' SLO.

Next to these developments, through an issue-centred approach, future work within the group will focus on the ways in which current societal trends impact organisational communication. In particular, this work will address the current epistemic crisis of trust, in which societal trust in organisations, politics, and institutions is declining (Edelman, 2023). In relation to this trend, future work addresses polarisation, in that organisations are asked to provide their perspectives on polarised sociopolitical issues (e.g., Eisele et al., 2022).

Decreasing societal and institutional trust provide new challenges to communication between organisations and their environment. With the aim to build and maintain relationships, organisational communication must become more concerned with establishing trust and meeting societal expectations. Accordingly, an information climate characterised by high levels of distrust adds an extra layer to corporate communication. Especially when this coincides with higher levels of (perceived) polarisation among different groups in society, this asks for new research to understand the challenges for organisations' interplay with their society that come with these trends. Such research can demonstrate how individual, organisational, and societal factors impact organisational communication on these issues and subsequent stakeholder perceptions (e.g., Burgers et al., 2023). Accordingly, future research will keep focusing on a related trend, namely the role of mis- and disinformation in issue arenas affecting organisations (e.g., Jin et al., 2020).

In addition, the ways in which changes in the media ecosystem affect organisations will be essential to consider when moving forward. Important trends we observe include the rise of new genres like political satire, which blends aspects of news, opinion, and entertainment (Brugman et al., 2021, 2022) and impacts audience perceptions on organisations and industries (Möller & Boukes, 2022). A second important trend is how fast developments with artificial intelligence (AI) and other technological innovations, including society's risk perceptions related to AI (Araujo et al., 2020), will change or challenge the ways both the media and organisations operate in the coming years. Such research will not only focus on how media shape perceptions of organisations' adoption of new technologies (e.g., Droog et al., 2020; Shaikh & Moran, 2022), but also how these technologies themselves impact the current media landscape and the behaviour by news media and organisations. In addition, future work will strive to further develop and refine important communication-scientific theories (e.g., agenda-setting theory, framing theory, news values theory) and to develop new (computational) methods (e.g., data donation) to further the study of organisations, media, and society (e.g., Araujo et al., 2022; Kroon et al., 2022).

A final important goal of ASCoR's Corporate Communication group is to not only strive towards academic impact, but also towards societal impact by making sure that results from our papers find their way to relevant (organisational) stakeholders. The group, for instance, also writes books aimed at professional audiences (e.g., Dekker et al., 2021), blog posts (e.g., Boukes, 2021), and columns (e.g., Burgers, 2023) about corporate communication research for general audiences. In addition, within ASCoR's Corporate Communication group opportunities are created to share research insights with professionals, such as the CSC Webinars and the CSC Insights Whitepaper Series developed through the Logeion Chair. The group strives to continue and expand these activities in the future.

Taken together, this work will not only advance our knowledge of communication-scientific theory in relation to organisations, but—hopefully—also provide organisations with important theory- and evidence-based insights to navigate today's complex and dynamic media landscape.

References

Araujo, T., Ausloos, J., van Atteveldt, W., Loecherbach, F., Moeller, J., Ohme, J., Trilling, D., van de Velde, B., de Vreese, C., & Welbers, K. (2022). OSD2F: An open-source data donation framework. *Computational Communication Research, 4*(2), 372–387. https://doi.org/10.5117/CCR2022.2.001.ARAU

Araujo, T., Helberger, N., Kruikemeier, S., & de Vreese, C. H. (2020). In AI we trust? Perceptions about automated decision-making by artificial intelligence. *AI & Society, 35*(3), 611–623. https://doi.org/10.1007/s00146-019-00931-w

Boukes, M. (2021, November 12). Hoe de regering burgers (niet) meekrijgt met de coronamaatregelen. *Stuk Rood Vlees.* https://stukroodvlees.nl/hoe-de-regering-burgers-niet-meekrijgt-met-de-coronamaatregelen/

Boukes, M., Jones, N. P., & Vliegenthart, R. (2022). Newsworthiness and story prominence: How the presence of news factors relates to upfront position and length of news stories. *Journalism, 23*(1), 98–116. https://doi.org/10.1177/1464884919899313

Boumans, J., Trilling, D., Vliegenthart, R., & Boomgaarden, H. (2018). The agency makes the (online) news world go round: The impact of news agency content on print and online news. *International Journal of Communication, 12*, 1768–1789. https://ijoc.org/index.php/ijoc/article/view/7109/2330

Brugman, B. C., Burgers, C., Beukeboom, C. J., & Konijn, E. A. (2021). From *The Daily Show* to *Last Week Tonight*: A quantitative analysis of discursive integration in satirical television news. *Journalism Studies, 22*(9), 1181–1199. https://doi.org/10.1080/1461670X.2021.1929416

Brugman, B. C., Burgers, C., Beukeboom, C. J., & Konijn, E. A. (2022). Satirical news from left to right: Discursive integration in written online satire. *Journalism, 23*(8), 1626–1644. https://doi.org/10.1177/1464884920979090

Burgers, C. (2023). Polarisatie. *C: Magazine voor communicatieprofessionals, 35*(4), 17.

Burgers, C., Beukeboom, C. J., Smith, P. A., & van Biemen, T. (2023). How live Twitter commentaries by professional sports clubs can reveal intergroup dynamics. *Computers in Human Behavior, 139*, Article 107528. https://doi.org/10.1016/j.chb.2022.107528

Deephouse, D. L., & Suchman, M. (2008). Legitimacy in organizational institutionalism. In R. Greenwood, C. Oliver, R. Suddaby, & K. Sahlin-Andersson (Eds.), *The SAGE handbook of organizational institutionalism* (pp. 49–77). Sage. https://doi.org/10.4135/9781849200387

Dekker, R., Jacobs, S. H. J., & Korthagen, I. (2021). *Openbaar bestuur en media.* Boom Bestuurskunde.

Droog, E., Burgers, C., & Kee, K. F. (2020). How journalists and experts metaphorically frame emerging information technologies: The case of cyberinfrastructure for big data. *Public Understanding of Science, 29*(8), 819–834. https://doi.org/10.1177/0963662520952542

Edelman. (2023). *2023 Edelman trust barometer: Global report.* https://www.edelman.com/sites/g/files/aatuss191/files/2023-01/2023%20Edelman%20Trust%20Barometer%20Global%20Report.pdf

Eisele, O., Tolochko, P., & Boomgaarden, H. G. (2022). How do executives communicate about crises? A framework for comparative analysis. *European Journal of Political Research, 61*(4), 952–972. https://doi.org/10.1111/1475-6765.12504

Etter, M., Ravasi, D., & Colleoni, E. (2019). Social media and the formation of organizational reputation. *Academy of Management Review, 44*(1), 28–52. https://doi.org/10.5465/amr.2014.0280

Fredriksson, M., & Pallas, J. (2021). Public sector communication and mediatization. In V. Luoma-aho & M. J. Canel (Eds.), *The handbook of public sector communication* (pp. 167–179). Wiley Blackwell. https://doi.org/10.1002/9781119263203.ch11

Gerken, F., & van der Meer, T. G. L. A. (2019). Crises frame dynamics: Frame diversity in news media and the role of governmental actors. *Journal of International Crisis and Risk Communication Research, 2*(2), 149–180. https://doi.org/10.30658/jicrcr.2.2.1

Ihlen, Ø., & Pallas, J. (2014). Mediatization of corporations. In K. Lundby (Ed.), *Handbook on mediatization of communication* (pp. 423–442). De Gruyter Mouton. https://doi.org/10.1515/9783110272215.423

Jacobs, S. H. J., & Liebrecht, C. C. (2023). Responding to online complaints in webcare by public organizations: The impact on continuance intention and reputation. *Journal of Communication Management, 27*(1), 1–20. https://doi.org/10.1108/JCOM-11-2021-0132

Jacobs, S. H. J., & Wonneberger, A. (2017). Did we make it to the news? Effects of actual and perceived media coverage on media orientations of communication professionals. *Public Relations Review, 43*(3), 547–559. https://doi.org/10.1016/j.pubrev.2017.03.010

Jacobs, S. H. J., & Wonneberger, A. (2019). Dealing with increasing complexity: Media orientations of communication managers in public sector organizations. *International Journal of Communication, 13*, 918–937. https://ijoc.org/index.php/ijoc/article/view/9571

Jin, Y., van der Meer, T. G. L. A., Lee, Y. I., & Lu, X. (2020). The effects of corrective communication and employee backup on the effectiveness of fighting crisis misinformation. *Public Relations Review, 46*(3), Article 101910. https://doi.org/10.1016/j.pubrev.2020.101910

Jonkman, J. G. F., Boukes, M., Vliegenthart R., & Verhoeven, P. (2020). Buffering negative news: Individual-level effects of company visibility, tone, and pre-existing attitudes on corporate reputation. *Mass Communication and Society, 23*(2), 272–296. https://doi.org/10.1080/15205436.2019.1694155

Kohring, M., Marcinkowski, F., Lindner, C., & Karis, S. (2013). Media orientation of German university decision makers and the executive influence of public relations. *Public Relations Review, 39*(3), 171–177. https://doi.org/10.1016/j.pubrev.2013.01.002

Kroon, A. C., & van der Meer, T. G. L. A. (2021). Who takes the lead? Investigating the reciprocal relationship between organizational and news agendas. *Communication Research, 48*(1), 51–76. https://doi.org/10.1177/0093650217751733

Kroon, A. C., van der Meer, T. G. L. A., & Vliegenthart, R. (2022). Beyond counting words: Assessing performance of dictionaries, supervised machine learning, and embeddings in topic and frame classification. *Computational Communication Research, 4*(2), 528–570. https://doi.org/10.5117/CCR2022.2.006.KROO

Luoma-aho, V., & Vos, M. (2010). Towards a more dynamic stakeholder model: Acknowledging multiple issue arenas. *Corporate Communications: An International Journal, 15*(3), 315–331. https://doi.org/10.1108/13563281011068159

Marschlich, S. (2022). *Corporate diplomacy: How multinational corporations gain organizational legitimacy: A neo-institutional public relations perspective.* Springer. https://doi.org/10.1007/978-3-658-36818-0

Marschlich, S., & Ingenhoff, D. (2022). The role of local news in constructing media legitimacy: How news media frames the sociopolitical efforts of multinational corporations in host countries. *Corporate Communications: An International Journal, 28*(7), 1–18. https://doi.org/10.1108/CCIJ-05-2022-0050

Möller, A. M., & Boukes, M. (2022). Satirizing the clothing industry on YouTube: How political satire and user comments jointly shape behavioral intentions. *Media Psychology, 25*(5), 724–739. https://doi.org/10.1080/15213269.2022.2066004

Nölleke, D., Scheu, A. M., & Birkner, T. (2021). The other side of mediatization: Expanding the concept to defensive strategies. *Communication Theory, 31*(4), 737–757. https://doi.org/10.1093/ct/qtaa011

Schafraad, P., & van Zoonen, W. (2020). Reconsidering churnalism: How news factors in corporate press releases influence how journalists treat these press releases after initial selection. *Communications, 45*(s1), 718–743. https://doi.org/10.1515/commun-2019-2075

Schafraad, P., van Zoonen, W., & Verhoeven, P. (2016). The news value of Dutch corporate press releases as predictor of corporate agenda building power. *Public Relations Review, 42*(3), 451–458. https://doi.org/10.1016/j.pubrev.2015.11.014

Schwinges, A., van der Meer, T. G. L. A., Lock, I., & Vliegenthart, R. (2022, May 26). Watchdog journalism in the age of Big Tech: How news media in the United States and Germany hold Big Tech corporations accountable [Paper presentation]. ICA Journalism Studies Division Graduate Student Colloquium, Paris, France.

Shaikh, S. J., & Moran, R. E. (2022). Recognize the bias? News media partisanship shapes the coverage of facial recognition technology in the United States. *New Media & Society.* Advance online publication. https://doi.org/10.1177/14614448221090916

Strauß, N., & Vliegenthart, R. (2017). Reciprocal influence? Investigating implicit frames in press releases and financial newspaper coverage during the German banking crisis. *Public Relations Review, 43*(2), 392–405. https://doi.org/10.1016/j.pubrev.2017.02.001

Strauß, N., Vliegenthart, R., & Verhoeven, P. (2016). Lagging behind? Emotions in newspaper articles and stock market prices in the Netherlands. *Public Relations Review, 42*(4), 548–555. https://doi.org/10.1016/j.pubrev.2016.03.010

Strauß, N., Vliegenthart, R., & Verhoeven, P. (2018). Intraday news trading: The reciprocal relationships between the stock market and economic news. *Communication Research, 45*(7), 1054–1077. https://doi.org/10.1177/0093650217705528

Strömbäck, J. (2008). Four phases of mediatization: An analysis of the mediatization of politics. *The International Journal of Press/Politics, 13*(3), 228–246. https://doi.org/10.1177/1940161208319097

van der Meer, T. G. L. A. (2016). *Communication in times of crisis: The interplay between the organization, news media, and the public* (Publication no. 9789463280389) [Doctoral dissertation, University of Amsterdam]. UvA-DARE. https://hdl.handle.net/11245/1.532222

van der Meer, T. G. L. A. (2018). Public frame building: The role of source usage in times of crisis. *Communication Research, 45*(6), 956–981. https://doi.org/10.1177/0093650216644027

van der Meer, T. G. L. A., & Jin, Y. (2020). Seeking formula for misinformation treatment in public health crises: The effects of corrective information type and source. *Health Communication, 35*(5), 560–575. https://doi.org/10.1080/10410236.2019.1573295

van der Meer, T. G. L. A., & Jonkman, J. G. F. (2021). Politicization of corporations and their environment: Corporations' social license to operate in a polarized and mediatized society. *Public Relations Review, 47*(1), Article 101988. https://doi.org/10.1016/j.pubrev.2020.101988

van der Meer, T. G. L. A., Kroon, A. C., Verhoeven, P., & Jonkman, J. (2019). Mediatization and the disproportionate attention to negative news: The case of airplane crashes. *Journalism Studies, 20*(6), 783–803. https://doi.org/10.1080/1461670X.2018.1423632

van der Meer, T. G. L. A., Kroon, A. C., & Vliegenthart, R. (2022). Do news media kill? How a biased news reality can overshadow real societal risks, the case of aviation and road traffic accidents. *Social Forces, 101*(1), 506–530. https://doi.org/10.1093/sf/soab114

van der Meer, T. G. L. A. & Verhoeven, P. (2013). Public framing organizational crisis situations: Social media versus news media. *Public Relations Review, 39*(3), 229–231. https://doi.org/10.1016/j.pubrev.2012.12.001

van der Meer, T. G. L. A., Verhoeven, P., Beentjes, H., & Vliegenthart, R. (2014). When frames align: The interplay between PR, news media, and the public in

times of crisis. *Public Relations Review, 40*(5), 751–761. https://doi.org/10.1016/j.pubrev.2014.07.008
van der Meer, T. G. L. A., Verhoeven, P., Beentjes, J. W. J., & Vliegenthart, R. (2017). Disrupting gatekeeping practices: Journalists' source selection in times of crisis. *Journalism, 18*(9), 1107–1124. https://doi.org/10.1177/1464884916648095
van der Meer, T. G. L. A., & Vliegenthart, R. (2018). The consequences of being on the agenda: The effect of media and public attention on firms' stock market performance. *Communications, 43*(1), 5–24. https://doi.org/10.1515/commun-2017-0027
Verhoeven, P. (2016). The co-production of business news and its effects: The corporate framing mediated-moderation model. *Public Relations Review, 42*(4), 501–521. https://doi.org/10.1016/j.pubrev.2016.03.006
Waeraas, A. (2020). Public sector communication and organizational legitimacy. In V. Luoma-aho & M. J. Canel (Eds.). *The handbook of public sector communication* (pp. 45–58). Wiley. https://doi.org/10.1002/9781119263203.ch2
Wonneberger, A., & Jacobs, S. (2016). Mass media orientation and external communication strategies: Exploring organizational differences. *International Journal of Strategic Communication, 10*(5), 368–386. https://doi.org/10.1080/1553118X.2016.1204613

About the authors

Toni G. L. A. van der Meer is an Associate Professor of Corporate Communication at ASCoR. His research broadly focuses on disruptions in communication. More specific topics relate to crisis communication (of corporations), processes of mediatisation, biases in news media coverage and selection, and the impact of misinformation. E-mail: G.L.A.vanderMeer@uva.nl

Christian Burgers is a Full Professor of Communication and Organisations and a Full Professor of Strategic Communication by special appointment (Logeion Chair) at ASCoR. Within the context of corporate communication, his research focuses on figurative framing, linguistic bias and stereotyping, satirical news, and misinformation. E-mail: C.F.Burgers@uva.nl

Sandra Jacobs is an Assistant Professor of Corporate Communication at ASCoR. Her research focuses on organisational legitimacy and reputation and analyses how (social) media and strategic communication facilitate and affect these constructions. In addition, she focuses on the mediated construction of issue arenas. E-mail: S.H.J.Jacobs@uva.nl

Pytrik Schafraad is a Lecturer in Corporate Communication and an Associate Member of ASCoR. His research focuses on the relationship between journalism and public relations and, more recently, also on diversity and inclusion communication. E-mail: P.H.J.Schafraad@uva.nl

Mark Boukes is an Associate Professor of Corporate Communication at ASCoR. He investigates the coverage and effects of economic news as well as infotainment formats (e.g., political satire). He is currently investigating the trust and financial challenges of the journalism industry. E-mail: M.Boukes@uva.nl

8. Contested Issues and Organisations: Media Debates about Sustainability and Diversity

Anke Wonneberger, Anne Kroon, Linda van den Heijkant, Christel van Eck, and Jeroen Jonkman

Abstract

Sustainability and diversity are two contested issues in contemporary societies that are often linked to organisations. This chapter outlines research by ASCoR's Corporate Communication group that assesses the causes, content, and consequences of media debates on sustainability and diversity. Organisations engage in these issue arenas in various roles: as employers having responsibilities toward internal stakeholders, as corporations seeking to respond to societal expectations to garner legitimacy, or as advocacy organisations entering public discourses with the aim of accelerating social change. Drawing on diverse methodological approaches with an emphasis on content-analytic techniques, the group's research contributes to communication theory and research on agenda setting, framing, climate change communication, and stereotype communication with relevant implications for communication practice.

Keywords: media debates, climate change, stereotypes, framing, agenda setting

Introduction

Responding to societal developments and expectations, organisations increasingly emphasise their environmental and social responsibilities. Engagement of organisations with environmental and social issues may

Wonneberger, A., Kroon, A., van den Heijkant, L., van Eck, C., & Jonkman, J. (2024). Contested issues and organisations: Media debates about sustainability and diversity. In T. Araujo & P. Neijens (Eds.), *Communication research into the digital society: Fundamental insights from the Amsterdam School of Communication Research* (pp. 139-154). Amsterdam University Press. https://doi.org/10.5117/9789048560592_ch08

positively affect organisational legitimacy and stakeholder relations and contribute to societal change (Stephan et al., 2016). Given the contested nature of such issues, organisational engagement can also trigger sharp criticism and controversy (Lock & Seele, 2017) and lead to organisational delegitimisation among stakeholders (Morsing & Schultz, 2006). Public discourses that can be observed in various media channels form issue arenas in which organisations and stakeholders interact through issues communication (Luoma-aho & Vos, 2010; van der Meer & Jonkman, 2021). Issues are contested if there are major disagreements among stakeholders concerning the nature or scope of the problem, responsibilities, and/or solutions. This may be reflected by polarisation around sustainability issues, scientific disagreement about sustainability solutions, or disagreement about the relevance of or solutions to diversity problems in society. Focusing on the role of organisations in such controversies, research by ASCoR's Corporate Communication group studies media debates about issues connected to sustainability and diversity. Following the issue-centred approach outlined in the previous chapter, this work contributes to unravelling the mediated interplay between organisations and their stakeholders in the context of sustainability and diversity issues (see Fig. 8.1) and to how these dynamics impact organisational legitimacy.

Stakeholders increasingly link environmental problems like climate change or plastic pollution to organisations, for instance, by holding corporations responsible for contributing to these problems or challenging them to contribute to sustainability transitions (Allen & Craig, 2016; Whelan et al., 2013). Such changes are often initiated or spurred by advocacy groups that contribute to multi-actor environmental and sustainability discourses in the news media and on social media platforms (Cox, 2010). In this context, research of the programme group assesses the roles and interplay of various organisations in sustainability issue arenas (Jacobs et al., 2021; Lock, Stachel, et al., 2020; Wonneberger & Vliegenthart, 2021).

In the context of social issues, recent years have been marked by an intensified debate about the role of prejudices and stereotypes about minorities or marginalised groups (Beukeboom & Burgers, 2019; Eberl et al., 2018). In addition to entering the public debate about diversity issues, organisations face challenges with respect to diversity, inclusion, and equity as employers. Members of the programme group study, for instance, how media representations of social groups, such as older workers, relate to stereotypical beliefs about job candidates during the hiring process as well as employees in the workplace (e.g., Kroon et al., 2018; Kroon et al., 2021; van Selm & van den Heijkant, 2021).

Figure 8.1. Conceptualising Issue Arenas on Contested Issues

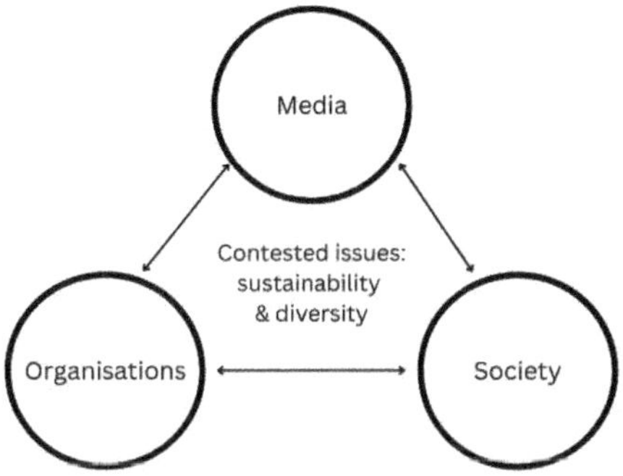

Empirical findings

The group's diverse methodological approach has enabled research on a broad spectrum of aspects related to sustainability and diversity discourses. ASCoR researchers have studied and categorised the *content* of frames, discourses, and portrayals in news coverage, on social media, and in organisational communications and investigated the *causes* of variation in this content utilising a mix of content- and network-analytic approaches (e.g., Hellsten et al., 2019; van den Heijkant et al., 2023a), with a specific focus on computational approaches (e.g., Kroon et al., 2020; van den Heijkant & Vliegenthart, 2018). We have tracked the *consequences* of this content on behavioural and attitudinal outcomes and identified potential *cures* by employing survey and experimental designs (e.g., Kroon et al., 2022; Wonneberger et al., 2020).

Sustainability

Research on the content and the causes of sustainability discourses contributes to a better understanding of fragmentation and polarisation of societal discourses on sustainability issues that are shaped by competing interests and ontologies and the forming of coalitions and counter-coalitions. In this context, socio-semantic network analysis allows disentangling how organisations and stakeholders are positioned and connected. Focusing on animal welfare issues, our studies have revealed how citizens play a central role in sustainability discourses on Twitter by forming socially and semantically distinct yet prominent and central clusters (Hellsten et al.,

2019; Wonneberger et al., 2021). Taking a bridging function, private social media users are of key strategic importance for organisations, for instance, to enhance the success of nonprofit campaigns (Wonneberger et al., 2021), or for public organisations which often take peripheral positions in discussion networks (Hellsten et al., 2019; Roth & Hellsten, 2022).

Zooming in on the content of social media debates showed that citizens often amplify aspects of discourses or organisational frames that are closest to their daily life experiences as consumers (Jacobs et al., 2021). But processes contributing to polarisation have also been identified. In the case of Twitter networks around the publication of an Intergovernmental Panel on Climate Change (IPCC) report engagement was specifically driven by a minority group of sceptical users (Roth & Hellsten, 2022).

In addition to mapping public discourses on sustainability issues, the group has studied related political and societal consequences. Studying the lawsuit of environmental organisation Urgenda against the Dutch state using time series analysis, we found that media attention for climate change litigation influenced political attention on this issue but also the media debate on climate policy and climate change in general (Wonneberger & Vliegenthart, 2021). A qualitative document analysis has been used combined with interviews to study framing strategies of lobbyists in a law case on nuclear emission data in Switzerland. This study revealed that the dominant frames mobilised by lobbyists were able to influence public policies (Lock, Stachel, et al., 2020).

In addition, media effects on the general public have been studied. Linking quantitative content analysis to survey data revealed how different climate change audience segments respond differently to news coverage about a climate change summit (Wonneberger et al., 2020). An online experiment found that individual attitudes and preferences of consumers play a role in legitimacy and credibility perceptions of corporate social responsibility (CSR) websites (Lock & Schulz-Knappe, 2018).

Potential *cures* identified by the group are oftentimes linked to shared underlying values. Taking a macro perspective, our research has identified distinct sustainability views for consumers, news media, and corporations which are, however, linked to similar values (Wonneberger et al., 2022). On the organisational level, a case study by Cornelissen et al. (2021) into the hybrid organisational identity of the Dutch social enterprise Fairphone showed that "rekeying"—which involves leaders transforming their previous understandings and framing of the organisational identity into a more dual interpretation—facilitated continuous adaptation. This enabled members within the organisation to improve their ability to integrate multiple sustainability objectives (economic, social, environmental) and values, resulting in a shared hybrid organisational identity.

Diversity

In today's society, diversity and inclusion have become increasingly important issues that organisations are expected to actively address. Within the Corporate Communication group, a line of research focuses on the portrayal of social groups in social, organisational, and news content, and the consequences thereof for how these social groups are perceived by organisational stakeholders and for their chances in the labour market. Particular attention has been given to older workers and older job seekers—a group that often encounters rigid stereotypes that limit their inclusion in the labour market and the workplace.

In research focusing on the content of diversity issues and the portrayal of social groups, manual content analyses revealed that older workers are typically portrayed in the news media in line with salient societal stereotypes. This portrayal includes both positive and negative elements: it emphasises warmth stereotypes by portraying older workers as trustworthy, while downplaying competence by depicting them as less adaptable and (technologically) capable (Kroon et al., 2018). These persistent mixed beliefs about older workers being warm but not competent are also reflected in organisational communication, particularly in job advertisements and recruitment practices (van Selm & van den Heijkant, 2021). In addition to studying the portrayal of older workers, the group has examined media portrayals of ethnic minorities. A large-scale automated content analysis of more than 3 million news articles revealed that Dutch news media tend to implicitly associate ethnic out-groups with low-status and high-threat stereotypes, especially, when compared to ethnic in-groups (Kroon et al., 2020; Kroon et al., 2021; Kroon & van der Meer, 2023).

Maintaining the content focus, additional research delved deeper into social policy debates in the Netherlands that pertain to older workers—focusing on the issues of employability (e.g., Kroon, van Selm, et al., 2017) and pension reform (e.g., van den Heijkant et al., 2023a). Focusing on how policy reforms are framed in the media showed that news media often problematise such issues and victimise older workers (Kroon, van Selm, et al., 2017). This tendency to emphasise problems rather than solutions when it comes to policy reforms is also evident in how news and social media frame the issue of raising the retirement age (van den Heijkant et al., 2023a).

To explore causes of variation in this framing, subsequent studies examined the social policy issues comparatively across different European countries, newspapers, and time periods. The findings indicate that frames and portrayals regarding the policy reform and the issue of older workers' employability are driven by differences in quoted sources, political orientation of news outlets, time frames, and political climate. For example, conservative news outlets are less likely to endorse measures that prolong active working lives beyond the formal retirement ages (Kroon, Vliegenthart, et al., 2017). This alignment of coverage with a newspaper's ideological orientation is also demonstrated in a study of the European news media's framing of the issue of pension reform, where left-leaning newspapers emphasised frames of inequality and negative consequences for citizens more than right-leaning newspapers (van den Heijkant et al., 2022).

Media portrayals and frames are not without consequences, as demonstrated by the group's research on the attitudinal and societal implications of media portrayals of issues related to the employability of older workers and the framing of retirement policies. Drawing on online experiments, this work showed that when the media present mixed stereotypes of older workers that positively emphasise warmth qualities but negatively emphasise competence qualities, this can activate implicit stereotypes and ultimately negatively influence the willingness to hire older workers (Kroon et al., 2016). Another study found that the news frames of pension reform (i.e., responsibility and justice frames) influenced how citizens perceived and responded to the proposed reform, suggesting an important role for news framing in shaping public opinion and informing the policymaking process (van den Heijkant et al., 2023b). Additional experimental research has provided evidence of how different forms of linguistic bias influence the creation and maintenance of social stereotypes in organisations (Beukeboom & Burgers, 2020; Burgers & Beukeboom, 2020).

Finally, to identify potential cures, several ways to counteract the harmful consequences of stereotypical media content were explored. We investigated raising awareness of implicit prejudice as a means to reducing self-selection of biased content. This approach is particularly effective for individuals with strong implicit but weak explicit biases—as these individuals are more receptive to such a strategy (Kroon et al., 2022). Another approach that has been studied to reduce age bias in organisations is *perspective-taking*, in which individual organisational actors actively consider the mental state of individuals who belong to the stereotypical group. A study by van Selm and van den Heijkant (2022) suggests that considering realistic facts about older workers, such as their low absenteeism and high levels of organisational citizenship, can reduce age bias when evaluating candidates' resumes in hiring processes.

Contributions to theory and practice

Jointly, this research provides rich insights linked to various communication theories. It shows, for instance, how agenda-setting processes are triggered by media coverage about climate advocacy or labour policies, how sustainability and diversity issues linked to organisations are framed on different media platforms, how communication can create and maintain social categories and stereotypes, and how stereotypes about minorities are presented in the media and with what effects.

Sustainability

While active and passive stakeholders had been discerned conceptually in research on issue arenas (Luoma-aho & Vos, 2010), the automated approach proposed by Hellsten et al. (2019) allows the empirical mapping of issue arenas as communication networks in which passive stakeholders get involved by active stakeholders. Integrating semantic and relational research traditions, this reveals how issue arenas are co-constituted by actors, their references to hashtags, other actors, and retweets (Wonneberger et al., 2021). Discerning structural and semantic positions in communication networks furthermore allows the identification of sub-discourses and opposing poles in discourses (Roth & Hellsten, 2022). Mapping issue arenas is of practical value for various social actors. Journalists who render complex controversies visible, benefit from advanced research on visualising mediated issues communication. This may indirectly contribute to opinion formation and decision-making by citizens and political actors. Moreover, these insights help organisations to anticipate issues and crisis communication focused on mitigating potential adverse effects and maintaining legitimacy (Strauß & Jonkman, 2017).

Sustainability research of the group has shown how news media take a moderating role in various strategic communication efforts. Studying climate change litigation as a public campaigning tool of climate activists, for instance, has revealed that media attention for a climate lawsuit against the Dutch government has influenced the political agenda. The identified bottom-up agenda-setting effects furthermore add to the literature on climate change-related agenda-setting effects. Moreover, this research has been acknowledged for confirming the thus far merely conceptualised indirect effects of climate change litigation beyond the direct effects of court rulings (IPCC, 2022; see Setzer & Vanhala, 2019). In addition, Lock, Stachel, et al. (2020) have developed a public affairs–specific integrated process model of framing. This model assigns a central role to news media for framing processes of lobbying activities by describing that frames displayed in the news in addition to strategic considerations influence lobbying organisations in taking over frames introduced by other actors (Lock, Stachel, et al., 2020).

This in addition to other studies have emphasised the importance of adequately identifying strategic target audiences and understanding how these might respond to different framing strategies. As several studies have highlighted the key role of the general public in disseminating organisational campaigning or framing efforts, it is important for organisations to better understand everyday life considerations of citizens related to an issue (Araujo & Kollat, 2018; Jacobs et al., 2021; Wonneberger et al., 2021). Segmenting audiences with respect to beliefs and attitudes about an issue is a useful approach to developing more clearly targeted communication strategies. While previous research had repeatedly identified distinct climate change audiences and linked these to different information needs (e.g., Metag et al., 2017), our research was the first to show segment-specific media effects in the context of climate change (Wonneberger et al., 2020).

Diversity

The line of research on diversity and inclusion has made significant theoretical contributions. Members of the group have developed the social categories and stereotype communication (SCSC) model (Beukeboom & Burgers, 2019). This communication-scientific theory explains the communicative and linguistic processes through which stereotypes are shared. The model proposes that stereotypes are built around three core cognitive variables: (1) perceived category entitativity (the degree to which a specific social category is seen as a unified group), (2) stereotype content (content of the specific cognitive perceptions people have of a social category) and (3) perceived category essentialism (the degree to which specific characteristics are perceived as inherent to its members and stable across time and social contexts). The SCSC model proposes that different forms of communicative bias contribute to the communication and maintenance of social stereotypes, related to (1) content and form of category labels, and

(2) which information is shared about social-category members in which ways. This model has helped to study the creation and maintenance of social stereotypes in organisations (Beukeboom & Burgers, 2020; Burgers & Beukeboom, 2020).

Expanding upon the content of stereotypes, the group has used the stereotype content model (SCM), which suggests that cognitive perceptions of social groups vary based on two dimensions: Competence (affecting perceived status) and warmth (affecting perceived threat) (e.g., Cuddy et al., 2008; Cuddy et al., 2009). This idea is supported by numerous psychological studies using experimental designs. The group's research extends the SCM to organisational, news, and social media and shows that diverse social groups are portrayed differently on these dimensions. This research finds that stereotypical media representations are often ambivalent, comprising a mix of positive and negative evaluations (Kroon et al., 2018). Moreover, implicit stereotypical beliefs have an impact on the type of content that individual audience members self-select, and news media can reinforce these beliefs (Kroon et al., 2021; Kroon et al., 2022). Our research highlights the crucial role of news media in shaping and reflecting stereotypical beliefs of various social groups, suggesting that media portrayals matter for the image problem of social groups, which can have tangible consequences, such as affecting the likelihood of hiring or promoting members of particular demographic groups (Kroon et al., 2016).

The group's empirical studies confirm and add to classical communication science theories, especially the theoretical underpinnings of framing. It sheds light on the relevance of framing theory in understanding social policy discussions surrounding employability and pension reform. A notable contribution is the tendency of social and news media to problematise such pressing social issues rather than propose solutions. By comparing media coverage in different European countries, our research makes a valuable contribution to understanding how contextual factors (such as the political orientation, time frames, and political climate of the news media) influence the framing of critical policy issues. By ultimately demonstrating the powerful influence of framing in shaping public opinion in this context, we add an essential piece to the puzzle of how social attitudes and policy discussions are shaped by the media and influence the potential (or lack thereof) for policy reform. The evidence from these contributions has been presented in several initiatives and discussions with governmental actors (e.g., the Netherlands Authority for the Financial Markets and the Ministry of Social Affairs and Employment) and organisational actors (e.g., AWVN, the largest employers' association).

Next steps

In this final section, we discuss how our ongoing and future research responds to societal developments with respect to sustainability and diversity. In doing so, ASCoR's Corporate Communication group will continue to engage in theory development and communication practice in these areas.

Sustainability

Debates about climate change have shifted from the existence and anthropogenic causes of climate change to proportionate climate solutions, for example, in related to the energy transition. Our sustainability research focus has shifted along toward understanding how organisations, the media, and stakeholders communicate in polarised issue arenas about climate solutions and sustainability transitions. Relying on communication science theory (Lock, Wonneberger, et al., 2020), such as agenda and frame building and setting, we continue to investigate the dynamics of communication processes on the macro level by analysing media discourses with mixed methods (e.g., Wonneberger et al., 2023). We are particularly interested in the mediated, issue-based interplay between corporations, environmental NGOs, political actors, scientists, and citizens and how this interplay may impact the legitimacy of involved organisations (Jonkman & Verhoeven, 2021). For example, we compare polarised media discourses across countries and investigate the agenda- and frame-building effects within polarised issue networks in the context of sustainable energy solutions. A new line of research focuses on interpersonal communication dynamics to understand how conversations between individuals with opposing views may result in starker climate change polarisation.

Understanding the drivers of climate change contestation and polarisation provides fertile ground for developing climate narratives to depolarise the debate. Hence, our future research aims at developing and testing such narratives for different messengers (communication professionals, climate scientists), target groups (Dutch citizens) (Jonkman et al., 2023), and about different messages (e.g., climate science) (van Eck, 2023). As psychological research increasingly points to the important role of values and emotions in how people shape their climate change risk perceptions (van Eck et al., 2020), we are particularly driven to test value-based and affective narratives.

Corporate Communication is well represented in the Environmental Communication Lab (EcoLab) of ASCoR, where new ideas for research are jointly developed. One of the goals is to put environmental and climate change communication science on the agenda of policymakers and practitioners. Members of the group are actively seeking involvement in new governmental initiatives where Dutch climate policy or communication campaigns are being developed. We further seek to bridge science and practice by regularly providing workshops to various audiences, ranging from climate scientists to communication professionals and the general public.

Diversity

Our group will continue to study how stereotypes and biases are constructed, reinforced, and potentially transformed. For instance, based on the SCSC model (Beukeboom & Burgers, 2019), organisational actors can become (more) aware of the linguistic means by which stereotypical views are formed, allowing them to develop interventions that can help change or prevent the formation of undesirable biases in an organisational context. Future research will focus on uncovering further mechanisms of linguistic bias and on studying linguistic bias in real-life settings.

For example, we will develop evidence-based tools to help employers recognise and mitigate biases in the hiring process and workplace. A new line of research will examine organisations' strategic communication on LGBTQ-related issues and related consequences for perceptions of authenticity and legitimacy. Furthermore, as digital technologies continue to expand, employers and recruiters have capitalised on these trends by using the fine-grained targeting settings of social media platforms to reach their desired audiences for job openings. Our research will evolve with these digital technologies, in particular, to examine concerns that the increased use of automated recruitment tools may be causing the delivery of job advertisements to be skewed across age groups, such that older individuals are disproportionately less likely to be targeted by employers, and, as a result, less likely to be exposed to job advertisements.

References

Allen, M. W., & Craig, C. A. (2016). Rethinking corporate social responsibility in the age of climate change: A communication perspective. *International Journal of Corporate Social Responsibility*, *1*(1), 1–11. https://doi.org/10.1186/s40991-016-0002-8

Araujo, T., & Kollat, J. (2018). Communicating effectively about CSR on Twitter: The power of engaging strategies and storytelling elements. *Internet Research*, *28*(2), 419–431. https://doi.org/10.1108/IntR-04-2017-0172

Beukeboom, C. J., & Burgers, C. (2019). How stereotypes are shared through language: A review and introduction of the social categories and stereotypes communication (SCSC) framework. *Review of Communication Research*, *7*, 1–37. https://doi.org/10.12840/issn.2255-4165.017

Beukeboom, C. J., & Burgers, C. (2020). Seeing bias in irony: How recipients infer speakers' stereotypes from their ironic remarks about social-category members. *Group Processes & Intergroup Relations*, *23*(7), 1085–1102. https://doi.org/10.1177/1368430219887439

Burgers, C., & Beukeboom, C. J. (2020). How language contributes to stereotype formation: Combined effects of label types and negation use in behavior descriptions. *Journal of Language and Social Psychology*, *39*(4), 438–456. https://doi.org/10.1177/0261927X20933320

Cornelissen, J. P., Akemu, O., Jonkman, J. G., & Werner, M. D. (2021). Building character: The formation of a hybrid organisational identity in a social enterprise. *Journal of Management Studies*, *58*(5), 1294–1330. https://doi.org/10.1111/joms.12640

Cox, R. (2010). *Environmental communication and the public sphere* (2nd ed.). Sage.

Cuddy, A. J. C., Fiske, S. T., & Glick, P. (2008). Warmth and competence as universal dimensions of social perception: The stereotype content model and the BIAS map. In M. P. Zanna (Ed.), *Advances in experimental social psychology* (Vol. 40, pp. 61–149). Elsevier. https://doi.org/10.1016/S0065-2601(07)00002-0

Cuddy, A. J. C., Fiske, S. T., Kwan, V. S. Y., Glick, P., Demoulin, S., Leyens, J.-P., Bond, M. H., Croizet, J.-C., Ellemers, N., Sleebos, E., Htun, T. T., Kim, H.-J., Maio, G., Perry, J., Petkova, K., Todorov, V., Rodríguez-Bailón, R., Morales, E., Moya, M., … Ziegler, R. (2009). Stereotype content model across cultures: Towards universal similarities and some differences. *British Journal of Social Psychology*, *48*(1), 1–33. https://doi.org/10.1348/014466608X314935

Eberl, J. M., Meltzer, C. E., Heidenreich, T., Herrero, B., Theorin, N., Lind, F., Berganza, R., Boomgaarden, H. G., Schemer, C., & Strömbäck, J. (2018). The European media discourse on immigration and its effects: A literature review. *Annals of the International Communication Association*, *42*(3), 207–223. https://doi.org/10.1080/23808985.2018.1497452

Hellsten, I., Jacobs, S., & Wonneberger, A. (2019). Active and passive stakeholders in issue arenas: A communication network approach to the bird flu debate on Twitter. *Public Relations Review*, *45*(1), 35–48. https://doi.org/10.1016/j.pubrev.2018.12.009

Intergovernmental Panel on Climate Change. (2022). *Climate change 2022: Mitigation of climate change. Contribution of Working Group III to the sixth assessment report of the Intergovernmental Panel on Climate Change*. https://doi.org/10.1017/9781009157926

Jacobs, S. H. J., Wonneberger, A., & Hellsten, I. R. (2021). Evaluating social countermarketing success: Resonance of framing strategies in online food quality debates. *Corporate Communications: An International Journal, 26*(1), 221–238. https://doi.org/10.1108/CCIJ-01-2020-0011

Jonkman, J. G. F., van Eck, C. W., Burgers, C. F. (2023, February 2–3). More or less optimistic? Citizen perspectives on the Dutch energy transition in five different audience segments [Paper presentation]. Etmaal van de Communicatiewetenschap, Enschede, the Netherlands.

Jonkman, J. G. F., & Verhoeven, P. (2021, September 6–9). Framing organizational issue legitimacy: Support and criticism in climate change news on Royal Dutch Shell [Paper presentation]. 8th European Communication Conference 2021.

Kroon, A. C., Trilling, D., & Raats, T. (2021). Guilty by association: Using word embeddings to measure ethnic stereotypes in news coverage. *Journalism & Mass Communication Quarterly, 98*(2), 451–477. https://doi.org/10.1177/1077699020932304

Kroon, A. C., Trilling, D., van der Meer, T. G. L. A., & Jonkman, J. G. F. (2020). Clouded reality: News representations of culturally close and distant ethnic outgroups. *Communications, 45*(s1), 744–764. https://doi.org/10.1515/commun-2019-2069

Kroon, A. C., & van der Meer, T. G. L. A. (2023). Who's to fear? Implicit sexual threat pre and post the "refugee crisis." *Journalism Practice, 17*(2), 319–335. https://doi.org/10.1080/17512786.2021.1916401

Kroon, A. C., van der Meer, T. G. L. A., & Pronk, T. (2022). Does information about bias attenuate selective exposure? The effects of implicit bias feedback on the selection of outgroup-rich news. *Human Communication Research, 48*(2), 346–373. https://doi.org/10.1093/hcr/hqac004

Kroon, A. C., van Selm, M., ter Hoeven, C. L., & Vliegenthart, R. (2016). Poles apart: The processing and consequences of mixed media stereotypes of older workers. *Journal of Communication, 66*(5), 811–833. https://doi.org/10.1111/jcom.12249

Kroon, A. C., van Selm, M., ter Hoeven, C. L., & Vliegenthart, R. (2017). Age at work: Explaining variation in frames of older employees in corporate and news media. *Journalism Studies, 18*(9), 1167–1186. https://doi.org/10.1080/1461670X.2015.1111162

Kroon, A. C., van Selm, M., ter Hoeven, C. L., & Vliegenthart, R. (2018). Reliable and unproductive? Stereotypes of older employees in corporate and news media. *Ageing and Society, 38*(1), 166–191. https://doi.org/10.1017/S0144686X16000982

Kroon, A. C., Vliegenthart, R., & van Selm, M. (2017). Between accommodating and activating: Framing policy reforms in response to workforce aging across Europe. *The International Journal of Press/Politics, 22*(3), 333–356. https://doi.org/10.1177/1940161217708525

Lock, I., & Schulz-Knappe, C. (2018). Credible corporate social responsibility (CSR) communication predicts legitimacy: Evidence from an experimental study. *Corporate Communications: An International Journal, 24*(1), 2–20. https://doi.org/10.1108/CCIJ-07-2018-0071

Lock, I., & Seele, P. (2017). Measuring credibility perceptions in CSR communication: A scale development to test readers' perceived credibility of CSR reports. *Management Communication Quarterly, 31*(4), 584–613. https://doi.org/10.1177/0893318917707592

Lock, I., Stachel, C., & Seele, P. (2020). Traveling frames: How corporate and civil society actors try to influence public administration and courts in a case on nuclear emission data in Switzerland. *Journal of Public Affairs, 20*(1), Article e2008. https://doi.org/10.1002/pa.2008

Lock, I., Wonneberger, A., Verhoeven, P., & Hellsten, I. R. (2020). Back to the roots? The applications of communication science theories in strategic communication research. *International Journal of Strategic Communication, 14*(1), 1–24. https://doi.org/10.1080/1553118X.2019.1666398

Luoma-aho, V., & Vos, M. (2010). Towards a more dynamic stakeholder model: Acknowledging multiple issue arenas. *Corporate Communications: An International Journal, 15*(3), 315–331. https://doi.org/10.1108/13563281011068159

Metag, J., Füchslin, T., & Schäfer, M. S. (2017). Global warming's five Germanys: A typology of Germans' views on climate change and patterns of media use and information. *Public Understanding of Science, 26*(4), 434–451. https://doi.org/10.1177/0963662515592558

Morsing, M., & Schultz, M. (2006). Corporate social responsibility communication: Stakeholder information, response and involvement strategies. *Business Ethics: A European Review, 15*(4), 323–338. https://doi.org/10.1111/j.1467-8608.2006.00460.x

Roth, C., & Hellsten, I. R. (2022). Socio-semantic configuration of an online conversation space: The case of Twitter users discussing the #IPCC reports. *Social Networks*. https://doi.org/10.1016/j.socnet.2022.06.007

Setzer, J., & Vanhala, L. C. (2019). Climate change litigation: A review of research on courts and litigants in climate governance. *WIREs Climate Change, 10*(3), Article e580. https://doi.org/10.1002/wcc.580

Stephan, U., Patterson, M., Kelly, C., & Mair, J. (2016). Organizations driving positive social change: A review and an integrative framework of change processes. *Journal of Management, 42*(5), 1250–1281. https://doi.org/10.1177/0149206316633268

Strauß, N., & Jonkman., J. G. F. (2017). The benefit of issue management: Anticipating crises in the digital age. *Journal of Communication Management, 21*(1), 34–50. https://doi.org/10.1108/jcom-05-2016-0033

van den Heijkant, L., Skovsgaard, M., & Vliegenthart, R. (2022, May 26–30). Framing the retirement age increase in Europe: A cross-national comparative study of frames in the news [Paper presentation]. The 72nd Annual ICA Conference, Paris, France.

van den Heijkant, L., van Selm, M., Hellsten, I. R., & Vliegenthart, R. (2023a). Framing pension reform in the news: Traditional versus social media. *Communications, 48*(2), 249–272. https://doi.org/10.1515/commun-2021-0058

van den Heijkant, L., van Selm, M., Hellsten, I. R., & Vliegenthart, R. (2023b). Justice matters: News framing effects on opinions about pension reform. *Work, Aging and Retirement.* Advance online publication. https://doi.org/10.1093/workar/waad011

van den Heijkant, L., & Vliegenthart, R. (2018). Implicit frames of CSR: The interplay between the news media, organizational PR, and the public. *Public Relations Review, 44*(5), 645–655. https://doi.org/10.1016/j.pubrev.2018.05.007

van der Meer, T. G. L. A., & Jonkman, J. G. F. (2021). Politicization of corporations and their environment: Corporations' social license to operate in a polarized and mediatized society. *Public Relations Review, 47*(1), Article 101988. https://doi.org/10.1016/j.pubrev.2020.101988

van Eck, C. W. (2023). The next generation of climate scientists as science communicators. *Public Understanding of Science.* Advanced online publication. https://doi.org/10.1177/09636625231176382

van Eck, C. W., Mulder, B. C., & van der Linden, S. (2020). Climate change risk perceptions of audiences in the climate change blogosphere. *Sustainability, 12*(19), Article 7990. https://doi.org/10.3390/su12197990

van Selm, M., & van den Heijkant, L. (2021). In search of the older worker: Framing job requirements in recruitment advertisements. *Work, Aging and Retirement, 7*(4), 288–302. https://doi.org/10.1093/workar/waaa026

van Selm, M., & van den Heijkant, L. (2022, May 26–30). Countering biases in employer communication towards older job seekers: Does perspective taking work? [Paper presentation]. The 72nd Annual ICA Conference, Paris, France.

Whelan, G., Moon, J., & Grant, B. (2013). Corporations and citizenship arenas in the age of social media. *Journal of Business Ethics, 118*(4), 777–790. https://doi.org/10.1007/s10551-013-1960-3

Wonneberger, A., Hellsten, I. R., & Jacobs, S. H. J. (2021). Hashtag activism and the configuration of counterpublics: Dutch animal welfare debates on Twitter. *Information, Communication & Society, 24*(12), 1694–1711. https://doi.org/10.1080/1369118X.2020.1720770

Wonneberger, A., Lock, I. J., & Steenbeek, P. (2022, May 25). Clashing views? Comparing sustainability understandings across Dutch news, corporations, and consumers [Paper presentation]. The 72nd Annual ICA Conference, Preconference "The Science of Science Communication: Mapping the Field," Paris, France.

Wonneberger, A., Meijers, M. H. C., & Schuck, A. R. T. (2020). Shifting public engagement: How media coverage of climate change conferences affects climate change audience segments. *Public Understanding of Science, 29*(2), 176–193. https://doi.org/10.1177/0963662519886474

Wonneberger, A., van Eck, C. W., Jonkman, J. G. F., & Burgers, C. (2023, February 2–3). From slaughtering trees to the complexities of renewable energies: Polarized biomass discourse in the Netherlands [Paper presentation]. Etmaal van de Communicatiewetenschap, Enschede, the Netherlands.

Wonneberger, A., & Vliegenthart, R. (2021). Agenda-setting effects of climate change litigation: Interrelations across issue levels, media, and politics in the case of Urgenda against the Dutch government. *Environmental Communication, 15*(5), 699–714. https://doi.org/10.1080/17524032.2021.1889633

About the authors

Anke Wonneberger is an Associate Professor of Corporate Communication at ASCoR and Co-Director of the Environmental Communication Lab (ECoLab). Her research interests include strategic communication of nonprofit organisations and environmental media discourses. E-mail: A.Wonneberger@uva.nl

Anne Kroon is an Associate Professor of Corporate Communication at ASCoR. She studies bias in digital media using computational techniques and experiments. Her research examines how social groups are portrayed in digital media and the resulting effects on prejudice and discrimination. E-mail: A.C.Kroon@uva.nl

Linda van den Heijkant is an Assistant Professor of Corporate Communication at ASCoR. Her research is dedicated to exploring controversial issues in the media and their influence on public attitudes. In addition, she studies media portrayals of social groups and their impact on prejudice in organisational contexts. E-mail: L.vandenHeijkant@uva.nl

Christel van Eck works as an Assistant Professor of Corporate Communication at ASCoR. Her research focuses on climate change communication, science communication, and polarisation dynamics. Being passionate about communicating her research beyond academia, she regularly gives workshops and presentations that bridge the science and practice of climate change communication. E-mail: C.W.vanEck@uva.nl

Jeroen Jonkman is an Assistant Professor of Corporate Communication at ASCoR. His research focuses on the organisational and societal implications of strategic communication in mediated multi-actor networks forming around sustainability issues. E-mail: J.G.F.Jonkman@uva.nl

9. Tailored Health Communication in a Digital World

Eline S. Smit, Annemiek J. Linn, Minh Hao Nguyen, Adriana Solovei, Melanie de Looper, and Julia C. M. van Weert

Abstract

Technology has gained an important role in health communication in recent decades. One example is the use of health technology for tailoring, i.e., using adapted, rather than generic, messages based on the receiver's unique characteristics and preferences. ASCoR researchers have investigated content tailoring, mode tailoring, and message frame tailoring, in the context of multiple topics (e.g., cancer treatment and smoking cessation), applying different methodologies. The results of these studies highlight the potential of tailoring health messages. In particular, tailoring of intervention content and delivery mode seem to be promising approaches to enhance effective health communication. Implications for practice and theory are discussed, as well as future directions, with technological developments high on our health communication research agenda.

Keywords: content tailoring, mode tailoring, message frame tailoring, customisation, health communication

Introduction

In the past decennia, healthcare has changed; from disease-centred to patient-centred, from single-disease-focused to multi-disease-focused, and from not only focused on treatment but also on prevention. These shifts highlight the diverse needs of the general population, patients, and clinicians. However, generic health communication has failed to meet individual needs and to address personal differences determining health (behaviour).

Smit, E. S., Linn, A. J., Nguyen, M. H., Solovei, A., de Looper, M., & van Weert, J. C. M. (2024). Tailored health communication in a digital world. In T. Araujo & P. Neijens (Eds.), *Communication research into the digital society: Fundamental insights from the Amsterdam School of Communication Research* (pp. 155-176). Amsterdam University Press. https://doi.org/10.5117/9789048560592_ch09

Although targeting (i.e., customising messages to shared characteristics of population subgroups) is an initial step towards making health communication personally relevant, this strategy is unable to accommodate all these personal differences. Tailoring, in contrast, refers to presenting health content in a manner that suits the preferences and characteristics of an individual person (Kreuter & Skinner, 2000). Tailored messages provide personalised information, stimulate self-referent encoding, are perceived as more personally relevant, are better processed, and are generally more effective than generic messages (Lustria et al., 2016).

Tailoring—further explained in the first part of this chapter—has been an important line of research in health communication at ASCoR. ASCoR researchers have focused on various aspects and types of tailoring, which will be further explained in the second part of this chapter, along with the main results that have been found. In the third part of the chapter, the implications of the work for health communication theory and practice are described. Finally, the chapter includes reflections regarding future directions of research in health communication.

Types of tailoring

Tailoring can be defined as "system-initiated tailoring" (data-driven personalisation performed by a computer system) or "user-initiated tailoring" (customisation or self-tailoring, through which users themselves can modify the form and/or content of the message) (Sundar & Marathe, 2010). System-initiated tailoring involves a computer system personalising a message based on data about socio-demographic characteristics, needs, and theoretically relevant factors, such as attitudes or self-efficacy (Lustria et al., 2016). Unlike this more traditional way of tailoring, user-initiated tailoring starts with the user and allows the user to customise the information or mode of delivery by selecting personally relevant elements.

In addition to possibilities for system-initiated or user-initiated tailoring, new digital technologies also offer new opportunities for what to tailor. The first type of tailoring is "content tailoring." Here, the content of the information is adapted to people's personal situation, preferences, and characteristics. This type of tailoring was originally (mostly) system-initiated. Content tailoring has been found to be (cost-)effective in promoting various health-related behaviours, but with small overall effect sizes, indicating room for further improvement (e.g., Krebs et al., 2010; Lustria et al., 2013). In the realm of digital messages, health communication scholars at ASCoR have ventured into mode tailoring and message frame tailoring. "Mode tailoring" involves adapting the delivery mode of a message or intervention to participants' learning styles or preferences (Smit, Linn, et al., 2015). The rationale for mode tailoring is based on the premise that individuals vary widely in their modality preferences (e.g., textual vs audiovisual information), processing styles (e.g., verbal vs visual learners), and cognitive motivations (e.g., low vs high need for cognition). For example, older people often must deal with age-related sensory (e.g., impaired vision and/or hearing) and cognitive (e.g., reduced processing speed) declines, which are likely to change their mode preferences for online information. Since message framing refers to the taking of a certain perspective when formulating a message, highlighting some bits of information while omitting others, we defined "message frame tailoring" as adjusting this perspective based on people's individual needs (Smit, Linn, et al., 2015). Although we theorise that an intervention's message frame can be tailored to different information-processing styles, such as respondents' need for cognition, need for affect, and need for autonomy, especially the latter has been the focus of several of our research projects (Altendorf et al., 2019; Altendorf, Hoving, et al., 2020; Altendorf, Smit, et al., 2020; Altendorf et al., 2022; Bol et al., 2019; Smit et al., 2019; van Strien-Knippenberg et al., 2022). Previous studies investigated whether interventions tailored to cognitive and affective needs are more effective than non-tailored interventions in changing health behaviour (intentions) (Williams-Piehota et al., 2003), but tailoring digital health communication to the need for autonomy seemed to have unexplored potential.

Empirical findings

ASCoR researchers have studied the effects of system- and user-initiated tailoring of the content, mode, and/or message frame of online health information in both experimental and real-life settings using technological advances and new tailoring processes. Experimental design is commonly used at ASCoR to evaluate and develop tailored health messages and to draw conclusions about cause-and-effect relationships. Randomised controlled trials (RCTs) are often used to draw conclusions about the effectiveness of tailored messages in a real-world clinical context, often informed by the Medical Research Council's (MRC) framework for the development and evaluation of complex interventions (Craig et al., 2008). Figure 9.1 gives an overview of the various studies on tailoring conducted at ASCoR. In the next sections, we will discuss our research on system-initiated and user-initiated tailoring of health information.

Figure 9.1. Overview of Tailoring Studies at ASCoR

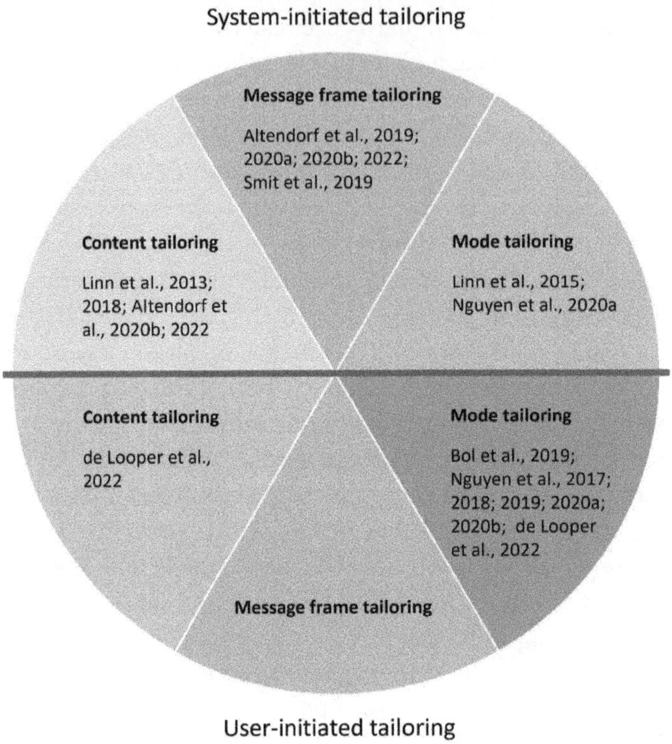

Findings on system-initiated tailoring

Below, we will discuss findings from our research on system-initiated tailoring relating to content, mode and/or message frame. We will first elaborate on the results from our experimental research, and then share the results from our research in real-world clinical contexts (via the RCTs).

Experimental research
The message frame tailoring studies in our line of research were always system-initiated. ASCoR researchers pioneered the use of the need for autonomy as a basis for tailoring and evaluating the message frames of online health behaviour change interventions. The concept of the need for autonomy is derived from self-determination theory (SDT) (Ryan & Deci, 2000), a theory that suggests that every person has a basic need for autonomy. Yet, people differ in the degree to which the need for autonomy is present. In several online message frame tailoring experiments, health messages

worded in an autonomy-supportive style (e.g., using words like "can" or "could" or giving choices, e.g., whether to read additional information) were compared with health messages worded in a more directive communication style (e.g., using words like "should" and "must" and not giving choices). People's individual need for autonomy was considered as a potential effect moderator in these studies (Altendorf et al., 2019; Altendorf, Smit, et al., 2020; Smit et al., 2019). Contrary to expectations, there was no moderating effect when testing a digital alcohol reduction intervention (Altendorf et al., 2019) and an online smoking cessation intervention (Altendorf, Smit, et al., 2020). However, in a vegetable intake intervention, providing choices—rather than the using of autonomy-supportive language—was found to be an easy-to-implement strategy to increase effectiveness, particularly for people with high autonomy needs (Smit et al., 2019).

In another online experiment, we tested if system-initiated tailoring to mode preference was more effective than non-tailoring in terms of recall and intention to perform breast self-examination. Breast self-examination instructions were either tailored to the participant's mode preference (text, text with illustrations, or video) or non-tailored by randomly providing one of these modes. The intention to perform breast self-examination was higher if the instruction was tailored to the mode preference. No effects on recall of information were found (Linn et al., 2015).

Randomised controlled trials

In one of our studies using the MRC framework, we systematically developed a tailored multimedia intervention (TMI) to improve medication adherence among chronically ill patients (Linn et al., 2013). The TMI combined content tailored technology with interpersonal communication. Technology was used to collect data on patients' barriers to adherence, to send tailored text messages for long-term support with minimal effort and to support clinicians in tailoring their communication to patients' needs during consultations. An RCT was then conducted in six hospitals. Results showed positive changes in the intervention group on important antecedents of medication adherence: patient satisfaction with nurses' affective communication and self-efficacy at treatment initiation. The effect on self-efficacy was maintained after six months (Linn et al., 2018).

To build on the results from the online experimental studies on message frame tailoring, we also considered the need for autonomy in real-world research. Instead of merely including the need for autonomy as a potential effect moderator, we tested the effectiveness of message frame tailoring based on the need for autonomy, in isolation and in combination with content tailoring, in an RCT with a 2 (message frame tailoring vs no message frame tailoring) by 2 (content tailoring vs no content tailoring) design among adult smokers intending to quit smoking (Altendorf, Hoving, et al., 2020). Results revealed that message frame tailoring based on the need for autonomy may be an effective addition to conventional content tailoring for people with a high need for autonomy. Yet, for people with a low need for autonomy, message frame tailoring was not effective in its current form. More research is needed to investigate what type of tailoring might benefit this group.

An advantage of conducting experiments in a naturalistic setting is the opportunity to include a trial-based economic evaluation, i.e., an economic evaluation study based on data collected alongside a (randomised controlled) trial, allowing to compare the costs of an intervention to its (health) benefits. Despite the challenges involved (Smit, de Vries, et al., 2015), such evaluations can provide valuable information for policymakers on resource allocation, especially when the results of the evaluation are inconclusive or insignificant in terms of effectiveness alone. To illustrate, while effectiveness results suggested that the combination of message frame tailoring and content tailoring was only beneficial for participants with a high need for autonomy, the combination of these two tailoring strategies was found to have a high potential for cost-effectiveness (increased smoking abstinence) and cost-utility (increased quality of life), for the entire sample studied, thus providing good value for money (Altendorf et al., 2022).

Findings on user-initiated tailoring

In addition to system-initiated tailoring, health information can also be tailored directly by the receivers, which we refer to as user-initiated tailoring. In the sections that follow, we will first discuss findings from our experimental research on user-initiated tailoring relating to content and mode, followed by results from our research in real-world clinical contexts (via the RCTs).

Experimental research

In an experimental study on the effects of user-initiated content tailoring, customisation of a mobile health app (operationalised as the possibility to add specific features of the app to a personal dashboard) led to higher intentions to engage in physical activity for respondents with a higher need for autonomy, but not for those with a lower need for autonomy (Bol et al., 2019).

As for user-initiated mode tailoring, we experimentally tested the effect of an informative website for patients with cancer that allowed younger and older adults to self-tailor the mode in which the information was presented. The mode-tailored condition (as compared to non-tailored conditions) improved attention to the website and, consequently, recall in older adults, but not in younger adults (Nguyen et al., 2017). For both younger and older adults, mode tailoring positively influenced satisfaction with the attractiveness and comprehensibility of the website, as compared to non-tailored conditions (Nguyen et al., 2018). Furthermore, a comparison between user-initiated and system-initiated mode tailoring showed that both strategies were more effective than non-tailored information, but depending on the outcome variable (i.e., attention, satisfaction, recall), health literacy level, and age of recipients, one strategy worked better than the other (Nguyen, Bol, & King, 2020). In the research that followed, we examined the underlying mechanisms that explain the effectiveness of mode tailoring (Nguyen, Bol, & Lustria, 2020). User-initiated mode tailoring showed an increase in perceived active control, resulting in higher perceived message relevance and website engagement, and reduced cognitive load. Positive indirect effects of mode tailoring through these mechanisms were found for both website attitude and information recall (Nguyen, Bol, & Lustria, 2020).

Randomised controlled trials

As a follow-up to the experimental mode tailoring studies described above, an RCT was conducted in an academic hospital aimed at newly diagnosed patients with cancer. Results showed the positive effects of exposure to an online educational intervention, which allowed them to self-tailor the intervention delivery mode, on satisfaction with the information and anxiety for younger patients (<65), but not for older ones (Nguyen et al., 2019).

Also, in a pilot RCT in four hospitals we tested the effectiveness of the Patient Navigator, systematically developed, MRC framework–guided website for patients newly diagnosed with cancer to prepare themselves for a consultation with a clinician by self-tailoring the content and the modality of the information. Even though the tailoring function of the tool was not frequently employed, usage of the Patient Navigator resulted in positive evaluations of the information, decreased anxiety after the consultation, and patients using fewer words during consultations, without impairing their satisfaction. These findings provide an opportunity to further personalise communication during consultation and respond to other patients' needs (de Looper et al., 2022).

Appendix 9.1 gives a summary of the characteristics and results of ASCoR studies on tailoring.

Contributions to theory and practice

Traditional tailoring research primarily focused on the effectiveness of system-initiated content tailoring and exploring its underlying mechanisms. Yet, our work goes beyond content tailoring by investigating the effects of mode and message frame tailoring, and both system-initiated and user-initiated types of tailoring. We confirmed the role of perceived relevance as a mediator but also identified new theoretical mechanisms and boundary conditions related to system-initiated mode and message frame tailoring, as well as user-initiated tailoring, such as perceived active control and cognitive load.

Theoretical insights

The elaboration likelihood model (ELM) (Petty & Cacioppo, 1986) is core to research on the effectiveness of tailored health communication (Kreuter & Wray, 2003). It suggests that factors such as relevant knowledge, individual differences in need for cognition, and the cognitive resources available to the individual influence the ability to process information. Motivation to engage with information (e.g., attention to a message) also influences information processing: When individuals perceive messages as personally relevant, they are more motivated to process information deeply, leading to stronger attitude changes. The ELM's proposed mechanisms, including perceived relevance and attention, have previously been confirmed as important mediators of the effectiveness of content tailored communication (Lustria et al., 2016). Similarly, ASCoR researchers have confirmed mediators proposed by the ELM (i.e., perceived relevance, engagement), but also identified why and under what circumstances (i.e., moderators) tailoring interventions are most effective, and the theoretical mechanisms (i.e., mediators) that explain the effects of different tailoring strategies (i.e., beyond content tailoring). For example, user-initiated mode tailoring increases the ability and motivation to process information by increasing recipients' sense of active control and reducing cognitive load. Furthermore, message frame tailoring, and content tailoring, as well as their interaction, was found to significantly predict the perceived relevance of smoking cessation messages, which consequently predicted participants' self-determined motivation. In turn, self-determined motivation positively influenced attitudes and self-efficacy to quit smoking, and self-efficacy predicted smoking abstinence (Altendorf, Hoving, et al., 2020).

Relevant to understanding mode tailoring effects, the cognitive theory of multimedia learning (CTML) (Mayer & Moreno, 2003) has often been used as a framework for designing multi-modal information and understanding its effects on learning (information processing and recall) (Mayer, 2002). In short, CTML suggests that combining multiple modes that complement each other (including verbal and visual representations, such as text with visuals, or visuals with narration) enhances learning (as compared to unimodal information) because information processing is cognitively less demanding, which may lead to deeper information processing. As cognitive resources are distributed across different processing systems, the combination of modes is expected to reduce overall cognitive load and thus facilitate information processing (Mayer, 2002). However, variability in a person's prior knowledge or working memory capacity can influence how information is being processed, and, therefore, the CTML's notion that deeper learning is facilitated by multi-modal information is not always guaranteed (Ayres, 2015). Based on our research at ASCoR, we propose that limitations of CTML can be accounted for by considering variability in individual preferences, needs and/or abilities for the mode of information presentation (i.e., tailoring). Concretely, mode tailoring could be used as a design principle to optimise multimedia information and learning, especially in contexts where audiences are likely to vary in information-processing preferences and abilities.

ASCoR researchers have used the self-determination theory (SDT) (Ryan & Deci, 2000) to explain the effects of message frame tailoring. SDT suggests that autonomous motivation, driven by meeting psychological needs, like the need for autonomy, plays an important role in the initiation and maintenance of health-related behaviour (change) and subsequent positive health outcomes (Ng et al., 2012). However, while all individuals have a basic need for autonomy, differences exist in how this need manifests (Smit & Bol, 2020). Some people prefer to choose their own path towards change, whereas others prefer to be guided by expert advice (Resnicow et al., 2008; Resnicow et al., 2014). *Message frame tailoring* seemed especially helpful for people reporting a high need for autonomy (Altendorf, Hoving, et al., 2020)—suggesting that tailoring health communication may be key. This recognition could extend SDT's theoretical assumptions.

Practical insights

ASCoR researchers have explicitly focused on the development and evaluation of tailored health communication interventions among populations that are often underserved, e.g., older people. Older adults might benefit from mode tailoring, since it may increase their attention and recall (Nguyen et al., 2017), as well as their satisfaction with the attractiveness and comprehensibility of a website (Nguyen et al., 2018). Yet, the ability to self-tailor the modality through which older adults receive information did not always have such positive effects (de Looper et al., 2022; Nguyen et al., 2019). This highlights that findings from general adult populations cannot directly be generalised to patients, and age is not the only factor that may make people vulnerable—factors like socio-economic status and health literacy should also be considered.

Relatively recent research on the "digital health divide," i.e., the tendency for digital health interventions to benefit disadvantaged populations less (Cornejo Müller et al., 2020), further illustrates this. Western et al. (2021), for example, concluded that digital interventions aimed at the promotion of physical activity are effective for adults of high, but not of low, socio-economic status. Moreover, Szinay et al. (2023) found that older people, those in lower-skilled jobs, and those living in rural areas might benefit less from mobile interventions than others. Also, vulnerable populations have been found to be less likely to start using digital health interventions (Chesser et al., 2016) and to engage with them (Szinay et al., 2023). As uptake and engagement with intervention content are important determinants of intervention effectiveness, it is important to investigate the reasons for the digital divide and to actively involve people from vulnerable target groups in intervention design, e.g., through co-creation, before assuming that findings from previous work in this area are applicable to them.

In line with findings from other scholars, (e.g., Smit et al., 2013; Stanczyk et al., 2014), the economic evaluation research conducted at ASCoR showed that the combination of message frame tailoring and content tailoring has potential for cost-effectiveness and cost-utility (Altendorf et al., 2022). This is an important finding, since—especially in healthcare—limited budgets are usually available to implement effective lifestyle interventions on a large scale, and (healthcare) policymakers need to make decisions on resource allocation, while aiming to produce most value for money. ASCoR researchers have suggested practical strategies to overcome the challenges that come with the economic evaluation of digital lifestyle interventions (Smit, de Vries, et al., 2015).

Our research shows that integrating technology with interpersonal communication can improve the delivery of tailored healthcare. First, offering online health information as a preparation tool before consultations enhances information processing. However, it should be recognised that not all patients, especially older people, prefer computer-based information sources. Second, incorporating mode tailoring into online health information can benefit patients by improving website evaluations, increasing information recall, and reducing anxiety. While it may be tempting to include all information modes on a website without tailoring, this approach is thus not recommended. Third, to integrate tailoring effectively, it is important to consider differences in information-processing styles, preferences, needs, and abilities of different patient groups. For now, providing textual information as the fixed, primary mode of intervention delivery is recommended, as text remains the preferred choice for many patients.

Next steps

Despite the various tailored health communication interventions that have been designed and tested, there is a constant need to improve and redefine these interventions in line with technological developments.

Artificial intelligence to inform tailoring message content

ASCoR researchers have begun to explore the convergence of technology and interpersonal communication in the healthcare setting to optimise tailoring (Sanders et al., 2020). AI methods such as machine learning (ML), and natural language processing (NLP) can be used to improve tailored information provision. As part of several interdisciplinary teams, ASCoR researchers are involved in tailoring research starting with (1) the development of rule-based expert systems that integrate screening tools and treatment guidelines, (2) the analysis of routinely collected "messy" data in general practitioners' electronic health records, and (3) the analysis of harmonised datasets from different patient cohorts to develop prediction models that allow the calculation of individual health risks. The information obtained from these methods will be integrated into a clinical decision support system (CDSS). A CDSS can link individual patient characteristics in the electronic health record to a computerised clinical knowledge base to generate patient-specific treatment recommendations to the healthcare provider. This information can then be used to improve patient–provider communication and shared decision-making during consultations (de Wildt et al., 2023; Westerbeek et al., 2023).

Furthermore, there has been a recent shift towards using conversational agents for health communication in domains such as mental health, physical activity, and sexual health. Conversational agents are systems that can simulate conversations with users through natural language such as written text or voice, permitting automated two-way communication between user and system (Balaji et al., 2022). ChatGPT is a recent example of a conversational agent, using AI and NLP to engage in human-like conversations with users. In our future research, we will identify barriers, facilitators, underlying factors, affordances, and regulatory and ethical implications that determine people's (e.g., general public, patients, clinicians) motivation to use AI-based health technologies such as conversational agents. Current projects focus already on encouraging engaging, long-term personalised interactions with conversational agents for health promotion (Balaji et al., 2022).

Delivering just-in-time tailored interventions

New technologies allow us to advance tailoring in other ways as well, not only by providing the right information, but also by providing this information at the right time. Just-in-time adaptive interventions (JITAIs) (Nahum-Shani et al., 2018) are specifically designed to account for daily variations in health-related cognitions and behaviours. The main advantages of JITAIs have been described as the potential to (1) tailor interventions to the needs of individual users in real time to deliver support at the most promising moment, (2) adapt to input data, (3) be system-triggered, (4) deliver goal-oriented interventions, and (5) allow for customisation according to user preferences (Villinger et al., 2022). Because of these characteristics, JITAIs may increase engagement with and effectiveness of health behaviour interventions as compared to more traditional tailored interventions. Within ASCoR, initial attempts have been made to investigate the effectiveness of a JITAI for physical activity (Vos et al., 2023), but as the evaluation of JITAIs comes with both theoretical and methodological challenges, more work is needed in this area.

References

Altendorf, M. B., Hoving, C., van Weert J. C. M., & Smit, E. S. (2020). Effectiveness of message frame-tailoring in a web-based smoking cessation program: Randomized controlled trial. *Journal of Medical Internet Research, 22*(4), Article e17251. https://doi.org/10.2196/17251

Altendorf, M. B., Smit, E. S., Azrout, R., Hoving, C., & van Weert, J. C. M. (2020). A smoker's choice? Identifying the most autonomy-supportive message frame in an online computer-tailored smoking cessation intervention. *Psychology & Health, 36*(5), 549–574. https://doi.org/10.1080/08870446.2020.1802457

Altendorf, M. B., van Weert, J. C. M., Hoving, C., & Smit, E. S. (2019). Should or could? Testing the use of autonomy-supportive language and the provision of choice in online computer-tailored alcohol reduction communication. *Digital Health, 5*. https://doi.org/10.1177/2055207619832767

Altendorf, M. B., van Weert, J. C. M., Hoving, C., & Smit, E. S. (2022). An economic evaluation of an online computer-tailored smoking cessation intervention that includes message frame-tailoring: A randomized controlled trial. *PLOS Digital Health, 1*(9), Article e0000094. https://doi.org/10.1371/journal.pdig.0000094

Ayres, P. (2015). State-of-the-art research into multimedia learning: A commentary on Mayer's handbook of multimedia learning. *Applied Cognitive Psychology, 29*(4), 631–636. https://doi.org/10.1002/acp.3142

Balaji, D., He, L., Giani, S., Bosse, T., Wiers, R., & de Bruijn, G. J. (2022). Effectiveness and acceptability of conversational agents for sexual health promotion: A systematic review and meta-analysis. *Sexual Health, 19*(5), 391–405. https://doi.org/10.1071/SH22016

Bol, N., Høie, N. M., Nguyen, M. H., & Smit, E. S. (2019). Customization in mobile health apps: Explaining effects on physical activity intentions by the need for autonomy. *Digital Health, 5*. https://doi.org/10.1177/2055207619888074

Chesser, A., Burke, A., Reyes, J., & Rohrberg, T. (2016). Navigating the digital divide: A systematic review of eHealth literacy in underserved populations in the United States. *Informatics for Health and Social Care, 41*(1), 1–19. https://doi.org/10.3109/17538157.2014.948171

Cornejo Müller, A., Wachtler, B., & Lampert, T. (2020). Digital divide—Social inequalities in the utilisation of digital healthcare. *Bundesgesundheitsblatt-Gesundheitsforschung Gesundheitsschutz, 63*(2), 185–191. https://doi.org/10.1007/s00103-019-03081-y

Craig, P., Dieppe, P., Macintyre, S., Michie, S., Nazareth, I., & Petticrew, M. (2008). Developing and evaluating complex interventions: The new Medical Research Council guidance. *BMJ, 337*, Article a1655. https://doi.org/10.1136/bmj.a1655

de Looper, M., Smets, E. M. A., Schouten, B. C., Bolle, S., Belgers, E. H. J., Eddes, E. H., Leijtens, J. W. A., & van Weert, J. C. M. (2022). The Patient Navigator: Can a systematically developed online health information tool improve patient participation and outcomes related to the consultation in older patients newly diagnosed with colorectal cancer? *BMC Cancer, 22*, Article 109. https://doi.org/10.1186/s12885-021-09096-6

de Wildt, K. K., van de Loo, B., Linn, A. J., Medlock, S. K., Groos, S. S., Ploegmakers, K. J., Seppala, L., Bosmans, J. E., Abu-Hanna, A., van Weert, J. C. M., van Schoor, N. M., & van der Velde, N. (2023). *Effects of a clinical decision support system and patient portal for preventing medication-related falls in older fallers: Protocol of a cluster randomized controlled trial with embedded process and economic evaluations (ADFICE_IT)* [Manuscript submitted for publication].

Krebs, P., Prochaska, J. O., & Rossi, J. S. (2010). A meta-analysis of computer-tailored interventions for health behaviour change. *Preventive Medicine, 51*(3–4), 214–221. https://doi.org/10.1016/j.ypmed.2010.06.004

Kreuter, M. W., & Skinner, C. S. (2000). Tailoring: What's in a name? *Health Education Research, 15*(1), 1–4. https://doi.org/10.1093/her/15.1.1

Kreuter, M. W., & Wray, R. J. (2003). Tailored and targeted health communication: Strategies for enhancing information relevance. *American Journal of Health Behaviour, 27*(1), S227–S232. https://doi.org/10.5993/AJHB.27.1.s3.6

Linn, A. J., Alblas, M., van Weert, J., & Bol, N. (2015). Een kwestie van voorkeur? Een experimentele studie naar het aanpassen van gezondheidsinstructies aan de voorkeur voor modaliteit van de ontvanger. *Tijdschrift voor Communicatiewetenschap, 43*(4), 362–377. https://doi.org/10.5117/2015.043.004.362

Linn, A. J., van Dijk, L., van Weert, J. C., Gebeyehu, B. G., van Bodegraven, A. A., & Smit, E. G. (2018). Creating a synergy effect: A cluster randomized controlled trial testing the effect of a tailored multimedia intervention on patient outcomes. *Patient Education and Counseling, 101*(8), 1419–1426. https://doi.org/10.1016/j.pec.2018.03.017

Linn, A. J., van Weert, J. C., Smit, E. G., Perry, K., & van Dijk, L. (2013). 1 + 1 = 3? The systematic development of a theoretical and evidence-based tailored multimedia intervention to improve medication adherence. *Patient Education and Counseling, 93*(3), 381–388. https://doi.org/10.1016/j.pec.2013.03.009

Lustria, M. L. A., Cortese, J., Gerend, M. A., Schmitt, K., Kung, Y. M., & McLaughlin, C. (2016). A model of tailoring effects: A randomized controlled trial examining the mechanisms of tailoring in a web-based STD screening intervention. *Health Psychology, 35*(11), 1214–1224. https://doi.org/10.1037/hea0000399

Lustria, M. L. A., Noar, S. M., Cortese, J., van Stee, S. K., Glueckauf, R. L., & Lee, J. (2013). A meta-analysis of web-delivered tailored health behaviour change interventions. *Journal of Health Communication, 18*(9), 1039–1069. https://doi.org/10.1080/10810730.2013.768727

Mayer, R. E. (2002). Multimedia learning. *Psychology of Learning and Motivation, 41*, 85–139. https://doi.org/10.1016/S0079-7421(02)80005-6

Mayer, R. E., & Moreno, R. (2003). Nine ways to reduce cognitive load in multimedia learning. *Educational Psychologist, 38*(1), 43–52. https://doi.org/10.1207/S15326985EP3801_6

Nahum-Shani, I., Smith, S. N., Spring, B. J., Collins, L. M., Witkiewitz, K., Tewari, A., & Murphy, S. A. (2018). Just-in-time adaptive interventions (JITAIs) in mobile health: Key components and design principles for ongoing health behavior support. *Annals of Behavioral Medicine, 52*(6), 446–462. https://doi.org/10.1007/s12160-016-9830-8

Ng, J. Y. Y., Ntoumanis, N., Thøgersen-Ntoumani, C., Deci, E. L., Ryan, R. M., Duda, J. L., & Williams, G. C. (2012). Self-determination theory applied to health contexts: A meta-analysis. *Perspectives on Psychological Science, 7*(4), 325–340. https://doi.org/10.1177/1745691612447309

Nguyen, M. H., Bol, N., & King, A. J. (2020). Customisation versus personalisation of digital health information: Effects of mode tailoring on information processing outcomes. *European Journal of Health Communication, 1*(1), 30–54. https://doi.org/10.47368/ejhc.2020.003

Nguyen, M. H., Bol, N., & Lustria, M. L. A. (2020). Perceived active control over online health information: Underlying mechanisms of mode tailoring effects on website attitude and information recall. *Journal of Health Communication, 25*(4), 271–282. https://doi.org/10.1080/10810730.2020.1743797

Nguyen, M. H., Smets, E. M. A., Bol, N., Loos, E. F., van Laarhoven, H. W. M., Geijsen, D., van Berge Henegouwen, M. I., Tytgat, K. M. A. J., & van Weert, J. C. M. (2019). Tailored web-based information for younger and older cancer patients: Randomized controlled trial of a preparatory educational intervention on patient outcomes. *Journal of Medical Internet Research, 21*(10), Article e14407. https://doi.org/10.2196/14407

Nguyen, M. H., Smets, E. M. A., Bol, N., Loos, E. F., & van Weert, J. C. M. (2018). How tailoring the mode of information presentation influences older and younger adults' satisfaction with health websites. *Journal of Health Communication, 23*(2), 170–180. https://doi.org/10.1080/10810730.2017.1421729

Nguyen, M. H., van Weert, J. C. M., Bol, N., Loos, E. F., Tytgat, K. M. A. J., van de Ven, A. W. H., & Smets, E. M. A. (2017). Tailoring the mode of information presentation: Effects on younger and older adults' attention and recall of online information. *Human Communication Research, 43*(1), 102–126. https://doi.org/10.1111/hcre.12097

Petty, R. E., & Cacioppo, J. T. (1986). The elaboration likelihood model of persuasion. In R. E. Petty & J. T. Cacioppo (Eds.), *Communication and persuasion: Central and peripheral routes to attitude change* (pp. 1–24). Springer-Verlag. https://doi.org/10.1007/978-1-4612-4964-1_1

Resnicow, K., Davis, R. E., Zhang, G., Konkel, J., Strecher, V. J., Shaikh, A. R., Tolsma, D., Calvi, J., Alexander, G., Anderson, J. P., & Wiese, C. (2008). Tailoring a fruit and vegetable intervention on novel motivational constructs: Results of a randomized study. *Annals of Behavioural Medicine, 35*(2), 159–169. https://doi.org/10.1007/s12160-008-9028-9

Resnicow, K., Zhou, Y., Hawley, S., Jimbo, M., Ruffin, M. T., Davis, R. E., Shires, D., & Lafata, J. E. (2014). Communication preference moderates the effect of a tailored intervention to increase colorectal cancer screening among African Americans. *Patient Education and Counseling, 97*(3), 370–375. https://doi.org/10.1016/j.pec.2014.08.013y

Ryan, R. M., & Deci, E. L. (2000). Self-determination theory and the facilitation of intrinsic motivation, social development, and well-being. *The American Psychologist, 55*(1), 68–78. https://doi.org/10.1037/0003-066X.55.1.68

Sanders, R., Linn, A. J., Vliegenthart, R., van Eenbergen, M., van Weert, J. C. M., & Araujo, T. B. (2020). Patients' convergence of mass and interpersonal communication on an online forum: Hybrid methods analysis. *Journal of Medical Internet Research, 22*(10), Article e18303. https://doi.org/10.2196/18303

Smit, E. S., & Bol, N. (2020). From self-reliers to expert-dependents: Identifying classes based on health-related need for autonomy and need for external control among mobile users. *Media Psychology, 23*(3), 391–414. https://doi.org/10.1080/15213269.2019.1604235

Smit, E. S., de Vries, H., Oberjé, E. J. M., & Evers, S. M. A. A. (2015). Easier said than done: Overcoming challenges in the economic evaluation of internet-based lifestyle interventions. *The European Health Psychologist, 17*(1), 39–44. https://www.ehps.net/ehp/index.php/contents/article/view/764

Smit, E. S., Evers, S. M., de Vries, H., & Hoving, C. (2013). Cost-effectiveness and cost-utility of internet-based computer tailoring for smoking cessation. *Journal of Medical Internet Research, 15*(3), Article e57. https://doi.org/10.2196/jmir.2059

Smit, E. S., Linn, A. J., & van Weert, J. C. M. (2015). Taking online computer-tailoring forward: The potential of tailoring the message frame and delivery mode of online health behaviour change interventions. *The European Health Psychologist, 17*(1), 25–31. https://www.ehps.net/ehp/index.php/contents/article/view/762

Smit, E. S., Zeidler, C., Resnicow, K., & de Vries, H. (2019). Identifying the most autonomy-supportive message frame in digital health communication: A 2x2 between-subjects experiment. *Journal of Medical Internet Research, 21*(10), Article e1407. https://doi.org/10.2196/14074

Stanczyk, N. E., Smit, E. S., Schulz, D. N., de Vries, H., Bolman, C., Muris, J. W., & Evers, S. M. (2014). An economic evaluation of a video- and text-based computer-tailored intervention for smoking cessation: A cost-effectiveness and cost-utility analysis of a randomized controlled trial. *PloS ONE, 9*(10), Article e110117. https://doi.org/10.1371/journal.pone.0110117

Sundar, S. S., & Marathe, S. S. (2010). Personalization vs customization: The importance of agency, privacy and power usage. *Human Communication Research, 36*(3), 298–322. https://doi.org/10.1111/j.1468-2958.2010.01377.x

Szinay, D., Forbes, C. C., Busse, H., DeSmet, A., Smit, E. S., & König, L. M. (2023). Is the uptake, engagement, and effectiveness of exclusively mobile interventions for the promotion of weight-related behaviors equal for all? A systematic review. *Obesity Reviews, 24*(3), Article e13542. https://doi.org/10.1111/obr.13542

van Strien-Knippenberg, I., Altendorf, M. B., Hoving, C., van Weert, J. C. M., & Smit, E. S. (2022). Content and message frame tailoring in a web-based smoking cessation intervention: A study protocol for development and usability testing. *JMIR Formative Research*, *6*(4), Article e33886. https://doi.org/10.2196/33886

Villinger, K., Wahl, D. R., Debbeler, L. J., Koller, J. E., Brünecke, I., Lages, N. C., Schupp, H. T., & Renner, B. (2022). Using ecological momentary assessment to study variations in daily experiences and behaviors during the COVID-19 pandemic. *The European Health Psychologist*, *22*(4), 816–833. https://kops.uni-konstanz.de/server/api/core/bitstreams/d84d6c15-6d26-4373-992a-3b0f33f77984/content

Vos, A. L., de Bruijn, G. J., Klein, M. C., Lakerveld, J., Boerman, S. C., & Smit, E. G. (2023). SNapp, a tailored smartphone app intervention to promote walking in adults of low socioeconomic position: Development and qualitative pilot study. *JMIR Formative Research*, *7*, Article e40851. https://doi.org/10.2196/40851

Westerbeek, L., Dormosh, N., Heymans, M., Linn, A. J., Medlock, S., Schut, M., van der Velde, N., van Weert, H., van Weert, J. C. M., & Abu-Hanna, A. (2023). *SeNiors empOWred via big data to joint-manage their medication-related risk of falling in primary care: The SnowDrop project* [Manuscript submitted for publication].

Western, M. J., Armstrong, M. E., Islam, I., Morgan, K., Jones, U. F., & Kelson, M. J. (2021). The effectiveness of digital interventions for increasing physical activity in individuals of low socioeconomic status: A systematic review and meta-analysis. *International Journal of Behavioral Nutrition and Physical Activity*, *18*, Article 148. https://doi.org/10.1186/s12966-021-01218-4

Williams-Piehota, P., Schneider, T. R., Pizarro, J., Mowad, L., & Salovey, P. (2003). Matching health messages to information-processing styles: Need for cognition and mammography utilization. *Health Communication*, *15*(4), 375–392. https://doi.org/10.1207/S15327027HC1504_01

About the authors

Eline S. Smit is an Associate Professor of Health Communication at ASCoR. Her research interests concern digital health communication, including the integration of innovative digital health communication strategies (e.g., eHealth, mHealth) into the healthcare setting and novel computer-tailoring strategies (e.g., mode and message frame tailoring), and their application to physical and mental health-related behaviours. E-mail: E.S.Smit@uva.nl

Annemiek Linn is an Associate Professor of Health Communication at ASCoR. In her line of research she focuses on how technology and offline health communication are becoming intertwined. Her main interests are how the internet and patient–provider communication are converging and how new technologies can contribute to patient-centric care. E-mail: A.J.Linn@uva.nl

Minh Hao Nguyen is an Assistant Professor of Persuasion & New Media Technologies at ASCoR. Her research examines the content, uses, and consequences of digital media for health and well-being, and how digital inequalities play a role in this. She is currently working on multiple projects on digital well-being and digital disconnection. E-mail: M.H.Nguyen@uva.nl

Adriana Solovei is a Postdoctoral Researcher at ASCoR. Her current research focuses on the role of media use and interpersonal communication on public's adherence to preventive behaviours during (health) crises such as the COVID-19 pandemic. E-mail: A.Solovei@uva.nl

Melanie de Looper worked at ASCoR as a PhD candidate, lecturer, and postdoc researcher from 2016 to 2023. She is currently an Assistant Professor at Tilburg University. In her research, she focuses on digitalisation, online health (mis)information, and mental well-being, specifically among vulnerable populations (e.g., the Mhealth4all project). Furthermore, she investigates the role of mental health information on new social media in the well-being of youth (e.g., MPOWER). E-mail: m.delooper@tilburguniversity.edu

Julia van Weert is a Full Professor of Health Communication, Head of the Department of Communication Science, and Founding Director of the Amsterdam Center for Health Communication (ACHC). Her research unearths underlying mechanisms of online information processing and the impact of digital (AI-based) technologies on health outcomes of underprivileged and vulnerable people. E-mail: J.C.M.vanWeert@uva.nl

Appendix 9.1. Summary of tailoring studies at ASCoR

Authors	Type of tailoring	Study design	Target group	Main results
Altendorf et al. (2019)	System-initiated message frame tailoring	A 2 (autonomy-supportive language vs controlling language) x 2 (choice vs no choice) between-subjects experiment was conducted using an online computer-tailored alcohol reduction intervention.	Adult smokers	Neither autonomy-supportive nor controlling message frames had significant effects on perceived autonomy support or reactance, and there was no moderation from the need for autonomy.
Altendorf et al. (2020a)	System-initiated message frame tailoring	An online 2 (autonomy-supportive; controlling language) × 2 (choice; no choice) between-subjects design with the control condition (generic advice) with adult smokers intending to quit.	Adult smokers	Results revealed no significant effect of autonomy-supportive message frames on perceived autonomy-support or self-determined motivation, nor did the need for autonomy moderate these effects.
Smit et al. (2019)	System-initiated message frame tailoring	A web-based 2 (autonomy-supportive vs controlling language) × 2 (choice vs no choice) experiment was conducted.	Adults	Choice had a significant positive effect on the overall evaluation of the intervention, whereas for participants with a high need for autonomy, there was a significant positive effect on perceived relevance.
Linn et al. (2013; 2018)	System-initiated content tailoring	A cluster randomised controlled trial was conducted in six hospitals, eight nurses, and 160 chronic patients. Hospitals were randomised to either the experimental group (TMI intervention) or the control group (usual care)	Patients with inflammatory bowel disease	By combining tailored counselling with technology, this intervention resulted in positive changes in important prerequisites (i.e., satisfaction and self-efficacy) of medication adherence.
Linn et al. (2015)	System-initiated mode tailoring	An online experiment examining breast self-examination instructions that were either tailored to the participant's mode preference (text, text with illustrations, or video) or non-tailored by randomly providing one of these modes.	Adult women	The intention to perform breast self-examination was higher if the instruction was tailored to the mode preference. There were no effects of mode tailoring on the recall of information.

Authors	Type of tailoring	Study design	Target group	Main results
Nguyen et al. (2017)	User-initiated mode tailoring	Online experiment, with a 5 (condition: tailored vs text, text with illustrations, audiovisual, combination) × 2 (age: younger [25–45] vs older [≥ 65] adults) between-subjects design.	Adults	The mode-tailored condition (relative to non-tailored conditions) improved attention to the website and, consequently, recall in older adults, but not in younger adults. Younger adults recalled more from non-tailored information such as text only or text with illustrations, relative to tailored information.
Nguyen et al. (2018)	User-initiated mode tailoring	A 5 (condition: tailored vs text, text with visuals, text with audiovisual, combination) × 2 (age: younger [25–45] vs older [≥ 65] adults) online experimental study.	Adults	Mode tailoring positively influenced satisfaction with the attractiveness and comprehensibility of the website, as compared to non-tailored conditions. These effects on website satisfaction were not different for younger and older adults.
Nguyen et al. (2019)	User-initiated mode tailoring	A randomised controlled trial tested the effects of a web-based tailored educational intervention. The intervention group viewed a mode-tailored website (i.e., enabling patients to tailor information using textual, visual, and audiovisual modes) while the 3 control groups viewed a non-tailored website (i.e., text only, text with images, and text with videos).	Newly diagnosed younger (<65 years) and older (≥65 years) patients with cancer.	Younger patients viewing the mode-tailored website were more satisfied before consultation and reported lower anxiety after consultation. Mode tailoring yielded no other significant differences in patient outcomes.
Nguyen et al. (2020b)	User-initiated mode tailoring	Using a web-based experiment (tailoring vs no tailoring), this study investigated the underlying mechanisms that might explain the effects of *mode tailoring* (via textual, visual, and audiovisual information) on website attitudes and recall of online health information.	A representative sample of internet users (25–86 years)	Mode tailoring, enabling users to self-customise a health website's presentation mode increased users' perceived active control, which in turn contributed to higher perceived relevance and website engagement and reduced cognitive load. Positive indirect effects of mode tailoring through these mechanisms were found for both website attitude and information recall.

Authors	Type of tailoring	Study design	Target group	Main results
Bol et al. (2019)	User-initiated mode tailoring	A between-subjects (condition: mode-tailored vs standardised website) online experiment whereby study participants were exposed to a health-related app.	Adults	Results showed that customisation in mobile health apps does not increase perceived active control, autonomous motivation, or the intention to engage in physical activity. However, an interaction effect between customisation and the need for autonomy showed that customisation in mobile health apps leads to higher intentions to engage in physical activity for those with a greater need for autonomy, but not for those with a lesser need for autonomy.
de Looper et al. (2022)	User-initiated mode tailoring and content tailoring	A pilot randomised controlled study was conducted in the four participating hospitals.	Cancer patients	Only one-quarter of the patients visited the self-tailoring page and only three patients self-tailored the content or information presentation mode. Patients who received the Patient Navigator contributed less during the consultation by using fewer words than patients in the control condition and experienced less anxiety two days after the consultation than patients in the control condition.
Altendorf et al. (2020b)	System-initiated content and message-frame tailoring	A randomised controlled trial with a 2 (message frame-tailoring vs no message frame-tailoring) x 2 (content-tailoring vs no content-tailoring) design	Adult smokers intending to quit	Message frame tailoring based on the need for autonomy is an effective addition to conventional content-tailoring techniques in online smoking cessation interventions for people with a high need for autonomy; however, this is not effective in its current form for people with a low need for autonomy.

Authors	Type of tailoring	Study design	Target group	Main results
Altendorf et al. (2022)	System-initiated message frame tailoring and content tailoring	A randomised controlled trial using a 2 (message frame-tailoring, i.e., how messages are presented: autonomy-supportive vs controlling) x 2 (content-tailoring, i.e., what content is presented: tailored vs generic) design.	Adult smokers	The combination of message frame tailoring and content tailoring in online smoking cessation programmes seemed to have a high potential for cost-effectiveness (smoking abstinence) and cost utility (quality of life), thus providing good value for money.
Nguyen et al. (2020a)	System-initiated and user-initiated mode tailoring	In an online experiment participants viewed either a customised, personalised, or non-tailored (mismatched) website based on individual preferences for presentation mode. We analysed a 3 (condition) × 2 (health literacy level) × 2 (age group) between-subjects design, examining effects on: time spent online, attention, perceived relevance, website involvement, website satisfaction, and information recall.	Adults	Mode tailoring, by both customisation and personalisation, is more effective than no tailoring. However, contingent on the outcome variable (i.e., attention, website satisfaction, information recall), or health literacy level, and age group, different tailoring strategies show different effects.

10. The Importance of Consumer Empowerment in Dealing with Digital Persuasion

Edith G. Smit and Eva A. van Reijmersdal

Abstract

Within digital communication it has become more difficult for consumers to set boundaries for exposure to persuasion content and techniques. To empower consumers to critically deal with digital persuasion not only knowledge about persuasion tactics is needed, but also self-efficacy needs to be addressed to stimulate confidence in knowing how to control exposure to digital persuasion. This chapter combines several studies conducted within ASCoR's Persuasive Communication group to underline our approach in empowering consumers to deal with persuasion by providing knowledge and enhancing self-efficacy.

Keywords: disclosure, persuasion knowledge, self-efficacy, social media promotions, cookie-based advertising

Introduction

Within digital communication it has become more difficult for consumers to set boundaries for exposure to persuasion content and techniques. This has not only become more visible by what is offered, but also by new legislations to protect consumers online and applications to block advertising or to protect internet users from unwanted cookie-based advertising. Examples of increased persuasion techniques in digital communication are influencer marketing, advergames, social media promotions, or chatbots that promote recipe ingredients. Consumers are increasingly in need of

Smit, E. G., & van Reijmersdal, E. A. (2024). The importance of consumer empowerment in dealing with digital persuasion. In T. Araujo & P. Neijens (Eds.), *Communication research into the digital society: Fundamental insights from the Amsterdam School of Communication Research* (pp. 177-188). Amsterdam University Press. https://doi.org/10.5117/9789048560592_ch10

strategies to control their exposure to digital persuasion, making consumer empowerment more important than ever.

Empowerment is a widely used concept with variations in definitions (Pires et al., 2006). It refers to exerting some control and influence over decisions that affect one's life. Empowered individuals feel a sense of control, understand their sociopolitical environment, and are proactive in efforts to exert control (Amichai-Hamburger et al., 2008, originally described in Zimmerman, 1995). Empowerment in relation to persuasion means that consumers are skilled and feel in control in dealing with persuasion attempts, or at least deal with persuasion as a result of choice, and being aware of what they are dealing with.

This chapter focuses on two important ways to empower consumers to critically deal with digital persuasion: (1) enhancing knowledge about persuasion tactics that are used by advertisers, and (2) enhancing self-efficacy by knowing how to act.

Empirical findings

To study consumer empowerment, we used surveys and experiments and conducted (longitudinal) field studies. In this chapter, we combine several studies conducted within ASCoR's Persuasive Communication group to underline our approach in empowering consumers to deal with persuasion in the context of digital communication. We distinguish two ways to increase empowerment: empowerment by knowledge and empowerment by knowing how to act.

Empowerment by knowledge

One of the most important ways to empower consumers to deal with persuasion is to activate or enhance their knowledge about persuasion. Persuasion knowledge refers to the knowledge that people have about the intentions and tactics that are used in persuasive attempts (Friestad & Wright, 1994). In the digital media environment, advertising became more and more hidden, but also more sophisticated due to technologies such as artificial intelligence, tracking, and extended reality techniques, making it harder for people to recognise persuasion and to have a choice in being exposed to it.

Through disclosures or educational interventions, people can learn about when and how they were potentially influenced, and what the consequences are. An important part of our research and that of ASCoR colleagues focused on how disclosures can enhance knowledge and empower consumers. Disclosures are labels, pictograms, short texts, or spoken messages that inform the audience about the source and/ or persuasive tactics used in the message, such as sponsoring or cookie-based targeting.

In sponsored content the message of an advertiser is embedded in seemingly non-persuasive content in exchange for compensation (Eisend et al., 2020). Examples include influencer marketing where social media influencers present posts featuring brands, and advergames which are games created around the branded message of an advertiser (van Berlo et al., 2021). With sponsored content, it is not always clear for the audience who the sender of the message is (the influencer, the game developer, or the brand?) and what the intent is (entertainment, information, or persuasion?). Our research showed that disclosures that inform people about the sponsored nature of the content, or that communicate that the advertiser is the source, can empower consumers. Due to disclosures, they know that they are dealing with a persuasion attempt instead of neutral content. In several experiments focusing on various forms of sponsored content, such as advergames, influencer marketing posts on YouTube, Instagram, sponsored blogs, but also on brand placements in movies and television programmes, we and other ASCoR researchers showed that due to disclosures, adults were better able to recognise sponsored content as a form of advertising and to recognise its persuasive intent (i.e., their persuasion knowledge was activated) (for a meta-analysis, see Eisend et al., 2020).

We showed the empowering capacity of disclosures not only for adults (Breves et al., 2021), but also for minors. For example, our experiments showed that informing children and young adolescents about advertising tactics of YouTube influencers increased their persuasion knowledge (e.g., van Reijmersdal & van Dam, 2020; van Reijmersdal et al., 2020). More precisely, when adolescents between 12 and 16 years old watched an influencer video that was accompanied by a disclosure of advertising in the video, they were better able to recognise the advertising and to understand that the video had a persuasive intent (van Reijmersdal & van Dam, 2020).

In cookie-based advertising or targeted behavioural advertising, consumers encounter online advertising based on their previous internet behaviour or previously shared personal data (Boerman et al., 2017). EU regulation, in particular, the General Data Protection Regulation of 2018, aims to empower consumers to have control over their online data and protect their privacy. The rationale is that knowledge about technical and legal aspects of sharing personal data online would empower consumers. Indeed, our online privacy research shows that providing information about targeted advertising or the cookie warning icon increases knowledge about those topics (Smit et al., 2014; Strycharz et al., 2019; van Noort, et al., 2013). However, our research also shows that it matters what type of knowledge was provided (Strycharz et al., 2021). An experiment ($N = 294$) into the effects of technical versus legal knowledge in either a commercial or news website context showed that having legal knowledge about rights empowered consumers in their coping appraisal and motivation to reject online data collection (no differences between website contexts). However, technical knowledge about data flows only resulted in lower threat appraisal and potentially in more risk behaviour. Increased technical knowledge resulted in lowered perceived severity of the problem, indirectly making people less inclined to protect their privacy by turning personalisation off (Strycharz et al., 2019) or rejecting cookies (Strycharz et al., 2021).

Not only in the previously mentioned boomerang effect of providing technical knowledge in the context of personalised advertising, but also in the context of mobile apps such an opposite effect was demonstrated. Our study amongst 1,593 adults in the Netherlands showed that knowledge regarding the data collection and usage practices of mobile apps was limited. Moreover, this knowledge was not associated with more, but rather with less protection motivation and behaviour (Wottrich et al., 2019).

In addition, also in the field of sponsored content, knowledge of persuasion sometimes has a boomerang effect. A recent experiment among students, for example, showed that a disclosure of sponsorship in an online blog (versus no disclosure) led to increased perceptions of transparency which positively impacted persuasion in terms of more positive attitudes and higher purchase intentions (van Reijmersdal et al., 2023; see also Campbell & Evans, 2018; Krouwer et al., 2020). This effect is explained by people's appreciation of open and honest communication. Instead of tricking people into being persuaded without them knowing it, a disclosure is a transparent way of communicating, which can lead to sympathy and more openness to persuasion.

The good news is that in some cases, knowledge can really help to increase empowerment. Our research of the past decade on disclosures of sponsored content gave insights into some boundary conditions of empowerment by knowledge. With respect to the disclosure itself, in experiments among adults, we showed that the effectiveness of disclosures depends on their format in terms of wording (e.g., Boerman & Müller, 2022), timing (van Reijmersdal et al., 2020), and exposure duration (Boerman, et al., 2012). Summarising these studies, the following types of disclosures were shown to be effective in empowering adults and minors by means of knowledge: (1) more extensive and explicit disclosures that express not only sponsorship but also the intent of the sponsorship to increase sales and create positive associations; (2) forewarning disclosures that are shown before an online video starts instead of during the video; (3) longer disclosures that are shown for 6 instead of 3 seconds.

To conclude, disclosing persuasion by providing information may not always be enough to empower consumers. We have seen this in the domains of sponsored content, mobile apps, and targeted advertising. Only in those circumstances, when the disclosure is explicit and consumers are (cognitive) capable of dealing with the information, knowledge results in understanding what is happening and being able to make a choice (i.e., being empowered).

Empowerment by knowing how to act

Being empowered means being informed (knowledge) in such a way that acting upon it is possible, varying from doing nothing to knowing how to act and performing the actions required. Often there is a gap between what consumers know and what they do. This disassociation between knowledge on the one hand and lack of motivation or behaviour on the other hand is observed within different domains.

Several studies within the domain of sponsored content showed that whether disclosure could empower consumers, in the sense that their knowledge could be increased, depended on the motivation and ability of the consumer. For example, our meta-analysis, but also an experiment among adolescents, showed that disclosures are more effective among older children than among younger children (Eisend et al., 2020; van Reijmersdal & van Dam, 2020). That is, children lack the abilities to activate their knowledge, or do not have the knowledge that is necessary to deal with persuasion. An experiment also showed that disclosures of advergames had a stronger effect on persuasion knowledge when people were in a positive mood instead of a negative mood (van Reijmersdal et al., 2015). Also an experiment showed that when adults were first given a difficult task that depleted their cognitive resources, subsequent disclosures of television brand placement had less effect on their persuasion knowledge than when they were not depleted (Janssen et al., 2016).

Also an experiment on sponsored content in a movie showed that when levels of involvement with the content are high (van Reijmersdal, 2015) adults' persuasion knowledge was enhanced more than when they were low, implying higher motivation to process the disclosure and the (sponsored) content. Similarly, in our research into conversations with a chatbot (Ischen et al., 2022), we saw that only consumers who were more engaged in the conversation with the conversational agent (in other words, those who were willing to invest more cognitive resources into the interaction) were more likely to identify the persuasion attempts. This implies that not only informing, but also actively involving consumers, can help them to understand persuasive intents.

Another domain to counter this gap between knowledge, on the one hand, and lack of motivation or behaviour, on the other hand, is the domain of online privacy. In a recent study on online privacy protection, we anticipated that knowledge interventions may not always have the anticipated empowering effect (Boerman, Strycharz, et al., 2023). Based on the protection motivation theory (Rogers, 1975; Witte, 1992), we designed and successfully tested an intervention to train people to protect their privacy. This intervention resulted not only in increased self-efficacy, but also—and more importantly—in increased privacy protection behaviours in the short term (two weeks later) as well as in the long term (two months later). In other words, these results showed that it was possible to empower internet consumers by effectively training their behaviour to safeguard privacy and that this effect persisted over time.

It is important to address not only technical or legal knowledge, but also strategies to successfully react towards persuasion or act upon it. In other words, it is important to also empower consumers beyond knowledge via self-efficacy. Self-efficacy refers to one's own belief in the ability to succeed in a particular situation. Various studies in behaviour change have shown that people will engage in behaviour change once they perceive a reason of doing so (threat, concern) and more importantly when they believe they can make a difference in dealing with the threat (perceived efficacy) (for an overview of behaviour change models and references, see Smit et al., 2023). When people only perceive a threat, but do not believe that they are able to act effectively against the threat, in other words when they have weak efficacy beliefs, they are unlikely to change their behaviour.

Perceived efficacy beliefs are thus vital for change (Koletsou & Mancy, 2011). Not only in the context of sponsored content and online privacy protection, but also in the context of health and pro-environmental behaviour this belief is shown to be crucial. For instance, believing to be able to make a difference resulted in more adherence to policy measures to stop spreading COVID-19 (Meijers et al., 2021; Smit et al., 2023) or in more pro-environmental choices. Meijers et al. (2019), for instance, showed the potential of a visual impact metaphor at the point of action (i.e., the bin where clothes are collected). By visualising how recycling old cloths can give other persons new clothes impacted the attitudes and intentions of the participants by strengthening their response efficacy and increasing their persuasion knowledge. The Meijers et al. (2023) study confirmed how important efficacy beliefs are for stimulating pro-environmental actions, across different types of efficacy beliefs (personal and collective) and different types of behaviours, such as the intention to recycle more whenever possible, to donate to pro-environmental charity organisations, and using energy-efficient household equipment.

Digital tools might be specifically useful in this regard. For example, the study of Meijers et al. (2022) used virtual reality to educate the participants what the impact would be of choosing, for instance, a product that contains palm oil. Interestingly, these messages were not only able to influence choice immediately, but also affected purchase decisions two weeks later via strengthened self-efficacy beliefs.

In conclusion, when disclosing persuasion is not sufficient, interventions to show how to act and strengthen efficacy might help to encourage consumers to change their behaviour towards the promoted behaviour.

Contributions to theory and practice

Our studies have contributed to theory in several ways. First of all, we repeatedly showed that persuasion knowledge activation or enhancement is useful in empowering consumers to deal with digital persuasion. Also, our line of research showed that not only knowledge, but also evaluations of persuasion (i.e., perceptions of the content or tactic being unfair or providing biased information) play a crucial role in critically dealing with persuasion attempts. Our work led to refinement and extensions of the persuasion knowledge dimension by discerning both a cognitive and an attitudinal component (consisting of attitudes and critical evaluations of the persuasion) (Rozendaal et al., 2011; Boerman et al., 2018). Taken together, our research and that of ASCoR colleagues enabled us to better predict and understand the processes underlying consumer empowerment related to hidden digital persuasion by activating persuasion knowledge and resistance strategies.

Second, we provided insights into the boundary conditions of disclosure effects. Our work showed that consumer empowerment by knowledge is contingent upon the type of disclosure, but also upon individual factors. This calls for a more tailored approach to empowering consumers. Our work showed that it is important to take individual differences into account when analysing, predicting, and improving consumer knowledge about digital persuasion.

Our third contribution towards a better theoretical understanding of consumer empowerment is the need for focusing on the concept of self-efficacy. We showed that self-efficacy is crucial for effective consumer empowerment in the current digital media landscape especially since knowledge only is not enough in cases where the persuasion is more hidden, and consumers are overwhelmed and lack cognitive resources to make deliberate decisions.

Practically, our studies have contributed to interventions on disclosing forms of persuasion, such as online sponsored content and cookie-based targeted advertising. Our work and advice have helped to shape national and international legislations and regulations (e.g., The Social Code: YouTube, Dutch Media Law, Reclame Code Social Media & Influencer Marketing, FTC's Protecting Kids from Stealth Marketing, the cookie information sign) and resulted in discussions in the Dutch parliament on how to empower youth to deal with advertising. In addition, our research contributed to the visualisation of pro-environmental behaviour in a VR environment, helped the University of Amsterdam and its Faculty of Social and Behavioural Sciences to organise pro-environment waste bins, gave input for the Slimwonen app to help people to make energy-conscious choices within their household, and resulted in advice for the Dutch government on potential regulations concerning resistance to the wearing of bike helmets.

Next steps

The coming years will be used to further develop our understanding of persuasion and consumer empowerment. In particular, we see a need to advance our insights into the role of self-efficacy in empowering consumers. For instance, we will develop our intervention strategy to help people to overcome their feelings of privacy fatigue in dealing with online targeted advertising (Boerman, Strycharz, et al., 2023), and to understand persuasion tactics of conversational agents (Ischen et al., 2022). There is an urgent need to understand the balance of costs and benefits of various forms of dealing with digital persuasion from a consumer perspective within, for instance, the context of pro-environmental daily choices. Note that not all persuasion is misleading, dangerous, or bad. Many interventions aim to make the world healthier, happier, and green. In all cases it is important to do this in a transparent way. We will further study the development of digital applications to visualise daily choices in terms of energy (using mobile phone apps) or food purchases (with the help of VR applications and chatbot interactions) and make persuasion within those applications more transparent.

In addition, we will increasingly move away from a general approach to a more personalised approach focusing on empowering vulnerable groups. Our studies showed that vulnerable groups can be people with little persuasion knowledge in general (e.g., children or minors), people with little persuasion knowledge with respect to a specific medium or type of persuasion tactic (Boerman, van Reijmersdal, et al., 2023), or people with little knowledge on a certain topic (e.g., privacy, finance, or sustainability) (Breves & Liebers, 2022). We will study how these groups can be empowered to make informed decisions.

References

Amichai-Hamburger, Y., McKenna, K. Y. A., & Tal, S. A. (2008). E-empowerment: Empowerment by the internet. *Computers in Human Behavior, 24*(5), 1776–1789, https://doi.org/10.1016/j.chb.2008.02.002

Boerman, S. C., Kruikemeier, S., & Zuiderveen Borgesius, F. J. (2017). Online behavioral advertising: A literature review and research agenda. *Journal of Advertising, 46*(3), 363–376. https://doi.org/10.1080/00913367.2017.1339368

Boerman, S. C., & Müller, C. M. (2022). Understanding which cues people use to identify influencer marketing on Instagram: An eye tracking study and experiment. *International Journal of Advertising, 41*(1), 6–29. https://doi.org/10.1080/02650487.2021.1986256

Boerman, S. C., Strycharz, J., & Smit, E. G. (2023). How can we increase privacy protection behavior? A longitudinal experiment testing three intervention strategies. *Communication Research*. Advance online publication. https://doi.org/10.1177/00936502231177786

Boerman, S. C., van Reijmersdal, E. A., & Neijens, P. C. (2012). Sponsorship disclosure: Effects of duration on persuasion knowledge and brand responses. *Journal of Communication, 62*(6), 1047–1064. https://doi.org/10.1111/j.1460-2466.2012.01677.x

Boerman, S. C., van Reijmersdal, E. A., & Rozendaal, E. (2023). A closer investigation of dispositional persuasion knowledge of sponsored content: Exploring determinants and clusters. *Journal of Marketing Communications*. Advance online publication. https://doi.org/10.1080/13527266.2023.2175893

Boerman, S. C., van Reijmersdal, E. A., Rozendaal, E., & Dima, A. L. (2018). Development of the persuasion knowledge scales of sponsored content (PKS-SC). *International Journal of Advertising, 37*(5), 671–697. https://doi.org/10.1080/02650487.2018.1470485

Breves, P., Amrehn, J., Heidenreich, A., Liebers, N., & Schramm, H. (2021). Blind trust? The importance and interplay of parasocial relationships and advertising disclosures in explaining influencers' persuasive effects on their followers. *International Journal of Advertising, 40*(7), 1209–1229. https://doi.org/10.1080/02650487.2021.1881237

Breves, P., & Liebers, N. (2022). #Greenfluencing: The impact of parasocial relationships with social media influencers on advertising effectiveness and followers' pro-environmental intentions. *Environmental Communication, 16*(6), 773–787. https://doi.org/10.1080/17524032.2022.2109708

Campbell, C., & Evans, N. J. (2018). The role of a companion banner and sponsorship transparency in recognizing and evaluating article-style native advertising. *Journal of Interactive Marketing, 43*(1), 17–32. https://doi.org/10.1016/j.intmar.2018.02.002

Eisend, M., van Reijmersdal, E. A., Boerman, S. C., & Tarrahi, F. (2020). A meta-analysis of the effects of disclosing sponsored content. *Journal of Advertising, 49*(3), 344–366. https://doi.org/10.1080/00913367.2020.1765909

Friestad, M., & Wright, P. (1994). The persuasion knowledge model: How people cope with persuasion attempts. *Journal of Consumer Research, 21*(1), 1–31. https://doi.org/10.1086/209380

Ischen, C., Araujo, T. B., Voorveld, H. A. M., van Noort, G., & Smit, E. G. (2022). Is voice really persuasive? The influence of modality in virtual assistant interactions and two alternative explanations. *Internet Research, 32*(7), 402–425. https://doi.org/10.1108/INTR-03-2022-0160

Janssen, L., Fransen, M., Wulff, R., & van Reijmersdal, E. A. (2016). Brand placement disclosure effects on persuasion: The moderating role of consumer self-control. *Journal of Consumer Behavior, 15*(6), 503–515. https://doi.org/10.1002/cb.1587

Koletsou, A., & Mancy, R. (2011). Which efficacy constructs for large-scale social dilemma problems? Individual and collective forms of efficacy and outcome expectancies in the context of climate change mitigation. *Risk Management, 13*(4), 184–208. https://doi.org/10.1057/rm.2011.12

Krouwer, S., Poels, K., & Paulussen, S. (2020). Moving towards transparency for native advertisements on news websites: A test of more detailed disclosures. *International Journal of Advertising, 39*(1), 51–73. https://doi.org/10.1080/02650487.2019.1575107

Meijers, M. H. C., Remmelswaal, P., & Wonneberger, A. (2019). Using visual impact metaphors to stimulate environmentally friendly behavior: The roles of response efficacy and evaluative persuasion knowledge. *Environmental Communication, 13*(8), 995–1008. https://doi.org/10.1080/17524032.2018.1544160

Meijers, M. H. C., Scholtz, C., Torfadottír, R., Wonneberger, A., & Markov, M. (2021). Learning from the COVID-19 pandemic to combat climate change: Comparing drivers of individual action in global crises. *Journal of Environmental Studies and Sciences, 12*(2), 272–282. https://doi.org/10.1007/s13412-021-00727-9

Meijers, M. H. C., Smit, E. S., de Wildt, K., Karvonen, S., van der Plas, D., & van der Laan, L. N. (2022). Stimulating sustainable food choices using virtual reality: Taking an environmental vs health communication perspective on enhancing response efficacy beliefs. *Environmental Communication, 16*(1), 1–22. https://doi.org/10.1080/17524032.2021.1943700

Meijers, M. H. C., Wonneberger, A., Azrout, R., Torfadottír, R., & Brick, C. (2023). Introducing and testing the personal–collective–governmental efficacy typology: How personal, collective, and governmental efficacy subtypes are associated with differential environmental actions. *Journal of Environmental Psychology, 85*, Article 101915. https://doi.org/10.1016/j.jenvp.2022.101915

Pires, G. D., Stanton, J., & Rita, P. (2006). The internet, consumer empowerment and marketing strategies. *European Journal of Marketing, 40*(9/10), 936–949. https://doi.org/10.1108/03090560610680943

Rogers, R. W. (1975). A protection motivation theory of fear appeals and attitude change. *The Journal of Psychology, 91*(1), 93–114. https://doi.org/10.1080/00223980.1975.9915803

Rozendaal, E., Lapierre, M. A., van Reijmersdal, E. A., & Buijzen, M. (2011). Reconsidering advertising literacy as a defense against advertising effects. *Media Psychology, 14*(4), 333–354. https://doi.org/10.1080/15213269.2011.620540

Smit, E. G., Meijers, M. H. C., & Ischen, C. (2023). "Doing it together": Testing the impersonal impact hypothesis in the public health domain. *European Journal of Health Communication, 4*(3), 1–18. https://doi.org/10.47368/ejhc.2023.301

Smit, E. G., van Noort, G., & Voorveld, H. A. M. (2014). Understanding online behavioural advertising: User knowledge, privacy concerns and online coping behaviour in Europe. *Computers in Human Behavior, 32*, 15–22. https://doi.org/10.1016/j.chb.2013.11.008

Strycharz, J., Smit, E. G., Helberger, N., & van Noort, G. (2021). No to cookies: Empowering impact of technical and legal knowledge on rejecting tracking cookies. *Computers in Human Behavior, 120*, Article 106750. https://doi.org/10.1016/j.chb.2021.106750

Strycharz, J., van Noort, G., Smit, E., & Helberger, N. (2019). Protective behavior against personalized ads: Motivation to turn personalization off. *Cyberpsychology: Journal of Psychosocial Research on Cyberspace, 13*(2), Article 1. https://doi.org/10.5817/CP2019-2-1

van Berlo, Z. M. C., van Reijmersdal, E. A., & Eisend, M. (2021). The gamification of branded content: A meta-analysis of advergame effects. *Journal of Advertising, 50*(2), 179–196. https://doi.org/10.1080/00913367.2020.1858462

van Noort, G., Smit, E. G., & Voorveld, H. A. M. (2013). The online behavioural advertising icon: Two user studies. In S. Rosengren, M. Dahlen, & S. Okazaki (Eds.), *Advances in advertising research, vol. IV: The changing roles of advertising* (pp. 365–378). Springer Gabler. https://doi.org/10.1007/978-3-658-02365-2_28

van Reijmersdal, E. A. (2015). Disclosing brand placements in movies: Effects of disclosure type and movie involvement on attitudes. *Journal of Media Psychology, 28*(2), 78–87. https://doi.org/10.1027/1864-1105/a000158

van Reijmersdal, E. A., Brussee, E., Evans, N., & Wojdynski, B. W. (2023). Disclosure-driven recognition of native advertising: A test of two competing mechanisms.

Journal of Interactive Advertising, 23(2), 85–97. https://doi.org/10.1080/15252019.2022.2146991

van Reijmersdal, E. A., Lammers, N., Rozendaal, E., & Buijzen, M. (2015). Disclosing the persuasive nature of advergames: Moderation effects of mood on brand responses via persuasion knowledge. *International Journal of Advertising, 34*(1), 70–84. https://doi.org/10.1080/02650487.2014.993795

van Reijmersdal, E. A., Rozendaal, E., Hudders, L., Vanwesenbeeck, I., Cauberghe, V., & van Berlo, Z. M. C. (2020). Effects of disclosing influencer marketing in videos: An eye tracking study among children in early adolescence. *Journal of Interactive Marketing, 49*(1), 94–106. https://doi.org/10.1016/j.intmar.2019.09.001

van Reijmersdal, E. A., & van Dam, S. (2020). How age and disclosures of sponsored influencer videos affect adolescents' knowledge of persuasion and persuasion. *Journal of Youth and Adolescence, 49*(7), 1531–1544. https://doi.org/10.1007/s10964-019-01191-z

Witte, K. (1992). Putting the fear back into fear appeals: The extended parallel process model. *Communication Monographs, 59*(4), 329–349. https://doi.org/10.1080/03637759209376276

Wottrich, V. M., van Reijmersdal, E. A., & Smit, E. G. (2019). App users unwittingly in the spotlight: A model of privacy protection in mobile apps. *Journal of Consumer Affairs, 53*(3), 1056–1083. https://doi.org/10.1111/joca.12218

Zimmerman, M. A. (1995). Psychological empowerment: Issues and illustrations. *American Journal of Community Psychology, 23*(5), 581–599, https://doi.org/10.1007/BF02506983

About the authors

Edith G. Smit is a Full Professor and holds the Chair in Persuasive Communication and Empowerment at ASCoR. She studied persuasion strategies in medical television programmes, how low(er) literate people understand complex health information, green packaging cues, consumer attention to advertising while media multitasking, and effects of personalised communication and privacy concerns. E-mail: E.G.Smit@uva.nl

Eva A. van Reijmersdal is an Associate Professor of Persuasive Communication at ASCoR. She examines the persuasiveness of various forms of sponsored content, including influencer marketing and advergames. She also studies how children and adults can be empowered in dealing with sponsored content by using disclosures. E-mail: E.A.vanReijmersdal@uva.nl

11. Persuasion in an Algorithmic Context

Guda van Noort, Hilde Voorveld, and Joanna Strycharz

Abstract
The centrality of data and algorithms in media have changed the creation, delivery, and consumption of online content, including persuasive content. This chapter focuses on the algorithmic persuasion process between consumers and brands, and introduces it through the research conducted at ASCoR's Persuasion and New Media Technologies group. Algorithmic persuasion involves deliberate persuasive attempts that are data-driven and mediated by algorithms. We introduce central findings related to data used for algorithmic persuasion, its effects on individuals and society, as well as individuals' understanding of such persuasive processes. The chapter concludes with a research agenda identifying four central themes related to theory development, mis- and disinformation, consumer vulnerabilities, and algorithmic persuasion in different contexts.

Keywords: online advertising, data-driven communication, personalised communication, social media platforms, PKM

Introduction

In our digital society, media landscapes have become data-driven and algorithm-mediated: algorithms are used to automatically and dynamically create and deliver the content media users see. Algorithms have transformed online environments into persuasive architectures (Susser, 2019) that influence media users through processes that are subtle and unobtrusive. As such, our digital society caused a fundamental change in how persuasive communication is created, delivered, and consumed.

To this end, a key research theme in ASCoR's Persuasion & New Media Technologies group is algorithmic persuasion, which stems from traditional

van Noort, G., Voorveld, H., & Strycharz, J. (2024). Persuasion in an algorithmic context. In T. Araujo & P. Neijens (Eds.), *Communication research into the digital society: Fundamental insights from the Amsterdam School of Communication Research* (pp. 189-204). Amsterdam University Press. https://doi.org/10.5117/9789048560592_ch11

advertising but has changed due to the central role of consumer data and algorithms. Algorithmic persuasion is defined as the process of a deliberate attempt by a persuader to influence the beliefs, attitudes, and behaviours of people through online communication that is mediated by algorithms (Zarouali, Boerman, et al., 2022).

Our definition of algorithmic persuasion reflects the broad scope of the most recent definitions of advertising. Whereas earlier definitions of advertising focused exclusively on mass media: "A paid, mediated form of communications from an identifiable source designed to persuading the recipient to take action, now or in the future" (Richards & Curran, 2002, p. 74), advertising is currently defined as "[b]rand-initiated communication intent on impacting people" (Dahlen & Rosengren, 2016, p. 334). This definition addresses the fact that brands are increasingly using proprietary channels, such as social media and apps that are neither mediated nor paid, that it is sometimes hard to identify who initiated the communication (e.g., the brand or an influencer), and that "receivers" of communication are no longer passive recipients, but active, engaged users of media (Voorveld et al., 2018).

In comparison to advertising that is not mediated by algorithms, algorithmic persuasion (for example, on social media platforms like Instagram or TikTok) has unique elements that, from a consumer perspective, makes this type of advertising potentially more problematic and, therefore, warrants our efforts to investigate this phenomenon. Persuasion was traditionally used to target people at a group level, but algorithms make it possible to automatically target and personalise messages to individuals in a far more fine-grained way. So instead of having one TV spot and targeting it to a certain audience that is expected to be watching TV at that moment, it is now possible to develop a thousand different (tailored) social media ads and serve them to individuals who are most likely to respond based on their actual behaviour or even their psychological profile or current state.

In the remainder, we purposely refer to the concept as algorithmic "persuasion," and not as algorithmic "advertising," as we believe that other communication fields, such as corporate or political communication also use data, algorithms, tools and infrastructures to persuade individuals (such as online social media platforms) (see, e.g., Helberger et al., 2021). Therefore, the knowledge on how algorithmic persuasion works in the domain of advertising can—and needs to—be applied and tested in other communication fields as well. At the same time, persuasive communication for commercial purposes has its own legal framework (aimed at empowering consumers to make autonomous and informed decisions) and ethical implications. Therefore, the distinction between commercial communication and other communication fields is important. Hence, in this chapter, we focus on algorithmic persuasion in the communication context of consumers and brands.

The aim of this chapter is to discuss the empirical and theoretical knowledge on algorithmic persuasion with a focus on the domain of interactions between brands and consumers, based on research conducted by ASCoR researchers and in ASCoR's Persuasion & New Media Technologies group, in particular. We do so along the lines of an input–process–output model building on our group's past research to reflect on theoretical and methodological developments in the field. We also provide avenues for future research.

Empirical findings

We believe that algorithmic persuasion can best be understood via an input–process–output model. ASCoR researchers have introduced the algorithmic persuasion framework (APF) which consists of five conceptual components: input, algorithm, persuasion attempt, persuasion process, and persuasion effects (see Figure 11.1) (Zarouali, Boerman, et al., 2022).

In short, the framework addresses how data variables serve as input for different algorithmic techniques and algorithmic objectives, which influences the manifestations of algorithm-mediated persuasion attempts, informing how such attempts are processed and their intended and unintended persuasive effects. Unique about algorithmic persuasion is that it is a circular process. Via "feedback loops" the effects of a particular "persuasion attempt" becomes new input for future persuasion attempts that ultimately reinforces or alters algorithmic systems. As a simple example, imagine that a gambling brand has collected data on their consumers including on how they spend prize money, and that this brand develops a TikTok video or an Instagram ad for their online gambling services. It decides, based on their consumer profiles, to create different versions communicating the different possible ways to spend the prize money. Next, these data are combined with data and algorithms of the advertising platform (e.g., TikTok or Meta) and is used to deliver the ad with the specific ad claims to different consumer groups so that the claims are in line with the consumer profiles. Then, platform users through their interactions with the ads provide data that can serve as input to further optimise this persuasive strategy: their clicks, viewing time, and visits

Figure 11.1. The Algorithmic Persuasion Framework

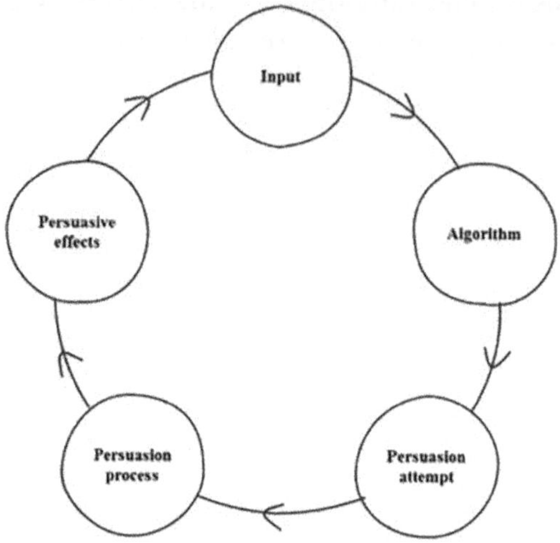

to the gambling website provide insight in which consumer (groups) are susceptible to which ad claim, which allows the further refinement of consumer profiles for future persuasion attempts.

Below, we use the APF as a conceptual overview to summarise our research. An extensive discussion of the model itself and how it can be used to test specific hypotheses can be found in Zarouali, Boerman, et al. (2022). Generally, a combination of traditional communication science methods as well as more innovative research methods such as different types of (big) data that are offered by new media technologies are used.

The key input variable in algorithmic persuasion is data. A common distinction made is between zero- (voluntarily provided by consumers), first- (collected by the brand/advertiser), second- (collected by another party in the persuasion process, such as a platform), and third-party data (e.g., collected by a party that is not involved in the persuasion process, such as a data broker) (see, e.g., Yun et al., 2020). Other distinctions are also made, such as between explicit data (i.e., wittingly disclosed by online media user) and implicit data (i.e., compilation and/or inferences from data). Implicit data, for instance, refer to personality traits that can be inferred based on interactions on social media (likes and shares) and the language a person uses, or their emotional states. While such inferences are commonly used in the industry as input for algorithmic persuasion, they can also be applied in persuasion research. For example, Zarouali, Dobber, et al. (2022), in collaboration with the industry, applied such inference algorithms in their research and concluded that individuals are more strongly persuaded by ads that match their inferred personality traits.

All these data are used to create and deliver personalised persuasive content to individuals. We define personalisation as "the strategic creation, modification, and adaptation of content and distribution to optimise the fit with personal characteristics, interests, preferences, communication styles, and behaviors" (Bol et al., 2018, p. 373). ASCoR research has shown that personalising advertising content and delivery increases its effectiveness (e.g., Kruikemeier et al., 2016; Maslowska et al., 2016; Segijn & Voorveld, 2021; Walrave et al., 2018; Zarouali, Dobber, et al., 2020; Zhang, Voorveld, et al., 2023; for an overview, see Boerman et al., 2017), but at the same time consumers experience it as a privacy violation and see potential for algorithmic manipulation (Boerman et al., 2017; de Keyzer et al., 2022; Limpf & Voorveld, 2015; Segijn, Strycharz, et al., 2021; Strycharz et al., 2019; Zhang, Voorveld, et al., 2023).

While all these data are considered personal, media users do not perceive them as equally personal or privacy sensitive. This also depends on the type of data and the type of organisation with which the data is shared (Smit et al., 2014). More specifically, ads that use less individual-specific and private data (such as browsing behaviour) are perceived to be more acceptable than ads that use more individual-specific and private information (such as name and email content) (Boerman et al., 2021).

Algorithms and persuasion techniques are another key component in algorithmic persuasion, next to data, and they represent the process through which data is used to influence individuals. A lot has been written about the black box nature of algorithms (e.g., Kitchin, 2017). For the domain of interactions between brands and consumers, at least three aspects related to algorithms' hidden nature are vital. Algorithms challenge people's awareness or knowledge about (1) the fact that they are being persuaded, because the commercial intent is often unclear, (2) how they are being persuaded, because people rarely have any knowledge of the persuasive practices and algorithms employed, and (3) by whom they are being persuaded, because algorithmic persuasion reflects a form of human–machine communication which makes it difficult for individuals to discern the source of a persuasion attempt (e.g., for a conceptual discussion on source orientation in human–chatbot communication, see van der Goot, 2022). Empirical research done outside ASCoR has shown that only a limited part of the population is aware of algorithms and understands what is happening (Eslami et al., 2015; Gran et al., 2021). Survey research at our group has indeed shown a low level of understanding on the use of data and algorithms in persuasion (Smit et al., 2014; Voorveld et al., 2023), while innovative online tracking of how individuals control use of their data for persuasion demonstrated that once aware, the majority of consumers opt-out from being exposed to such persuasion (Strycharz et al., 2019).

However, awareness is only a small part of the literacy that is needed for consumers to be able to effectively cope with algorithmic persuasion according to the persuasion knowledge model (PKM) (Friestadt & Wright, 1994), critical evaluations and effective coping skills are needed, too. Recent empirical work showed that different groups of people can be identified with regards to their cognitive understanding of algorithmic persuasion, affective beliefs about it, and coping mechanisms (Voorveld et al., 2023): the "Control Paradox," "Fatigued," "Unwitting but Critical," and "Skilled and Critical."

Perceptions with regard to data and algorithms logically influence processing and effects of algorithmic persuasion on individual consumers. When consumers realise that their data have been used to create or deliver an ad, the ad can be experienced as a surveillance episode, i.e., a directly observable instance of data collection (Strycharz & Segijn, 2022). For example, when an ad is closely related to a conversation one has had, it may make them think that the information on the conversation has been collected and analysed by advertisers (Frick et al., 2021). The use of algorithms for persuasive communication may make it more likely that individuals experience persuasion attempts as such a surveillance episode because algorithms make the personalisation of persuasive messages more fine-grained and advanced. Using qualitative research, ASCoR researchers have showed that the most common folk theory that consumers hold when it comes to surveillance is that commercial companies do everything for financial gains. At the same time, people have rather detailed ideas about the sources, purposes, and mechanisms involved (Zhang et al., 2021). These ideas are likely affecting the way consumers process an ad as negative reactions to data collection can spill over to advertising effects, negatively impacting ad effectiveness (Strycharz & Segijn, 2022).

Therefore, there is a new avenue of research at ASCoR that focuses on so-called surveillance responses, i.e., cognitive, affective, and behavioural responses to algorithmic persuasion that go beyond the ad itself. Regarding cognitive responses to data use for persuasion, research has shown that people have both positive and negative beliefs (Zhang et al., 2021). When it comes to consumer behaviour, data-driven advertising might lead to a situation in which consumers change their behaviour to control what behavioural information can be collected about them—so-called chilling effects (Büchi et al., 2022). In the context of advertising, chilling effects can, for example, include not visiting websites that collect data, or not using technology that is perceived as collecting data (e.g., smart devices). Experimental research at ASCoR has shown that data-driven advertising may lead to behavioural change, especially for consumers concerned about their privacy (Strycharz & Segijn, 2023). As chilling effects may undermine individual autonomy and well-being, research on the role of advertising in driving such behavioural change is highly societally relevant.

In sum, there is ample empirical research done to illustrate the different elements of the APF, but there is also a lot of interesting issues to explore in the future. We discuss these next steps, after summarising our contributions to theory and practice.

Contributions to theory and practice

ASCoR research provides theoretical models that guide the conceptualisation and research on persuasion in the context of continuous developments in new media technologies, and specifically in the context of algorithm-mediated communications. Next to the theoretical paper that introduced the APF that helps to conceptualise the phenomenon and that we discussed in the previous section (Zarouali, Boerman, et al., 2022), the automated brand-generated content (ABC) model (van Noort et al., 2020) helps to conceptualise algorithmic persuasion. The ABC model argues that not only consumer data but also brand data can be used as inputs for algorithms to serve short- and long-term persuasive goals. This model addresses the notion that in a data-driven marketing environment, the focus is on short-term effects (e.g., clicks, likes, and shares conversions), and the optimisation of content for each and every individual. In this optimalisation process, the way the brand is represented may vary from one consumer to the next. This is problematic for the brand as it dilutes brand identity, which consequently may jeopardise brand trust (and consumers who are less inclined to share their personal data). It may ignore or violate the communication of brand values that help relationship building, but not on long-term effects on brand perceptions. Therefore, this model proposes to balance the use of brand and consumer data, and long- and short-term impact goals of automated brand communication.

Lastly, literature reviews that were written on brand communication in social media (Voorveld, 2019), online behavioural advertising (Boerman et al., 2017), personalised communication (Eisend et al., 2023), and dataveillance

(Strycharz & Segijn, 2022) help to conceptualise the phenomenon of algorithmic persuasion and to understand consumers' responses towards it.

In our vision academic research not only makes theoretical contributions but ideally addresses and combats societal issues, too. Therefore, we actively contribute to society and practice by applying valorisation principles in our research. This means that we actively share our insights with societal partners and regulatory bodies, including the Authority for Consumers and Markets (ACM), and with communication professionals, for example, via the Dutch Foundation for Scientific Research on Commercial Communication (SWOCC). ASCoR researchers have good working contacts with different types of professional organisations, for example, platform media agencies, the Association of Advertisers (BvA), the Society for Advertising (GvR), and the Dutch Direct Marketing Association (DDMA). Some concrete examples in the past year: ASCoR researchers authored a policy report on vulnerable consumers and algorithms for the European Commission, they wrote a chapter for communication professionals on algorithmic persuasion in the *Marketingfacts Yearbook*, and they take part in the jury of an important advertising prize, the Advertiser's Jury of the Netherlands (Stichting Adverteerdersjury Nederland, SAN).

Next steps

Future research plans around the topic of algorithmic persuasion can be categorised in four different themes: Adaptation of existing persuasion theories, mis- and disinformation, rethink vulnerabilities, and algorithmic persuasion across domains and disciplines.

Adaption of existing persuasion theories

ASCoR researchers are currently reflecting on whether in the realm of algorithm-mediated communication we need new theories on persuasion processes between brands and consumers like the APF that helped us to conceptualise the phenomenon. For example, key theories on persuasion like the PKM (Friestadt & Wright, 1994) are challenged now algorithms are involved. According to this model, and other communication models, consumer knowledge on persuasion techniques "helps [people] identify how, when, and why marketers try to influence them" (p. 1), which helps them to cope with covert persuasion attempts like algorithmic persuasion. However, ASCoR research repeatedly showed that consumer knowledge on data collection does not result in empowerment per se (e.g., Strycharz et al., 2019). Also, algorithms are very complex, which makes it hard to communicate about it in a clear and understandable way (Senftleben et al., 2021).

Moreover, another reason why the PKM should potentially be updated is that it is not a dynamic model in which the key elements of the model "agent/source," "tactics," and "topic" influence and follow each other. According to the PKM, a person needs to have knowledge on agent/source, tactics, and topic and that this helps to cope with a persuasion attempt. Algorithms change the fundamental principle of this process. If algorithms register that a person likes (or ignores) a certain topic, algorithms can change the topic and tactics, and even sources/agents (i.e., which influencer or brand is shown in a person's feed) of the persuasive attempt (Voorveld, 2022). The traditional PKM does not account for these dynamics and, therefore, a new, adapted persuasion theory is needed (van Noort & Meppelink, 2021).

Mis- and disinformation

A prominent and pressing issue is that new technologies like AI and algorithms can be used to manipulate persuasive content. Mis- and disinformation in the context of algorithmic persuasion is present in the form of, for example, fake advertising. Such advertising can be defined as brand-related content spread by an individual or a brand pretending to be a different, genuine brand with the purpose of deceiving consumers. The questions arise: What do consumers need in order to cope with such deceptive attempts? and How can new media technology or other technologies such as blockchain technology—i.e., "a means for storing information and transactions in secure, decentralized manner" (Peres et al., 2023, p. 1)—empower consumers to distinguish fake from real information?

Rethink vulnerabilities

Unintended and unwanted consequences of (algorithm-mediated) persuasion is another pressing issue that ASCoR researchers are focusing on. In this respect a pressing issue is the need to rethink the concept of vulnerable audiences in relation to persuasive communication. A new definition of digital vulnerability that was developed in cooperation between our team members and academics at the University of Amsterdam's Institute for Information Law (IViR) is "a universal state of defencelessness and susceptibility to (the exploitation of) power imbalances that are the result of increasing automation of commerce, datafied consumer–seller relations, and the very architecture of digital marketplaces" (Helberger et al., 2022, p. 176). Traditionally, consumer groups like elderly and younger audiences are seen as most vulnerable, but in a digital and algorithm-driven context this is changing. The individualised and self-learning nature of algorithmic persuasion makes it possible to exploit the vulnerability of specific individuals in trying to influence them in an unprecedented and powerful way (Susser, 2019). These individuals do not necessarily belong to groups that have traditionally been defined as vulnerable. For persuasive communication scholars this brings a question of defining and empirically identifying individuals particularly vulnerable to algorithmic persuasion and the potential harms it can inflict.

Algorithmic persuasion across domains and disciplines

Furthermore, the omnipresence of algorithmic persuasion in all communication fields and in all media drives future research on algorithmic persuasion. For example, a pressing issue is to what extent processes and effect of algorithm-mediated persuasive communication can be generalised to other communication domains (including corporate, political, and health communication). This is important as it is known that current insights cannot simply be generalised across communication domains (e.g., health, politics, advertising). People differentiate between these communication contexts and, for example, make different trade-offs between privacy and the advantages of algorithm-mediated communication depending on the context (e.g., de Keyzer et al., 2022) and this influences the impact of algorithmic persuasion across domains. Hence, research is needed that embeds the effects of algorithmic persuasion on individuals and society in different communication domains. As findings show that consumers are not literate about algorithms in marketing communications (Voorveld et al., 2023) and that it is difficult to increase algorithmic literacy in this domain, the questions may arise (1) whether in political campaigning voters understand algorithmic processes, (2) should be empowered to recognise why they are exposed to political messages, and (3) should be legally protected against algorithmic persuasion just as consumers are. Given that for other domains, different laws and regulations apply, collaborating with other domains within the communication science discipline is needed. Also, extending existing collaborations with IViR is crucial here.

At the same time, we should focus on the potential of algorithmic persuasion to solve pressing societal issues, and to reduce or even bridge gaps within society. Such issues can be related to climate change, inequality, and health, including gambling and smoking cessation, problematic alcohol consumption, and being overweight. As an example: Algorithm-based recommender systems in the food domain (applied on food websites but also in apps of major online grocery stores recommending recipes to consumers) could be a useful tool to persuade people towards a more healthy food consumption pattern by identifying and overcoming barriers regarding healthy food choice, and recommending healthy food that are within the consumer's latitude of acceptance (Starke et al., 2023). Investigating the potential of algorithm-mediated persuasion to address pressing societal issues automatically means that it is necessary to collaborate with other disciplines and societal partners.

A final challenge is the constantly changing media landscape, fuelled by rapid technological developments. As a result, we constantly need to reflect on whether existing theory and knowledge fit these new developments in media technology. At the same time we should anticipate such developments in media technology and proactively inform society at large about possible intended and unintended persuasion processes and effects, even when such developments are still in their infancy. In this respect we believe the Internet of Things (IoT) is one of the major developments in media. Algorithms are part of all mediated communication, including websites, search engines, and social media, but also smart TVs, billboards, and so-called "smart" objects (e.g., the voice-driven Google Assistant or Amazon's Alexa). Such smart media and devices are on the rise and are increasingly connected to each other. Therefore, a final topic that will be on the agenda is how to understand algorithmic persuasion in a context of IoT, a world in which not just documents (images, text, information) are hyperlinked, and people are linked to each other (social media), but in which devices are also connected. In this context, the role that connected devices will play in persuasion and in our understanding of persuasive processes becomes central.

References

Boerman, S. C., Kruikemeier, S., & Bol, N. (2021). When is personalized advertising crossing personal boundaries? How type of information, data sharing, and personalized pricing influence consumer perceptions of personalized advertising. *Computers in Human Behavior Reports*, *4*, Article 100144. https://doi.org/10.1016/j.chbr.2021.100144

Boerman, S. C., Kruikemeier, S., & Zuiderveen Borgesius, F. J. (2017). Online behavioral advertising: A literature review and research agenda. *Journal of Advertising*, *46*(3), 363–376. https://doi.org/10.1080/00913367.2017.1339368

Bol, N., Dienlin, T., Kruikemeier, S., Sax, M., Boerman, S. C., Strycharz, J., Helberger, N., & de Vreese, C. H. (2018). Understanding the effects of personalization as a privacy calculus: Analyzing self-disclosure across health, news, and commerce contexts. *Journal of Computer-Mediated Communication*, *23*(6), 370–388. https://doi.org/10.1093/jcmc/zmy020

Büchi, M., Festic, N., & Latzer, M. (2022). The chilling effects of digital dataveillance: A theoretical model and an empirical research agenda. *Big Data & Society*, *9*(1). https://doi.org/10.1177/20539517211065368

Dahlen, M., & Rosengren, S. (2016). If advertising won't die, what will it be? Toward a working definition of advertising. *Journal of Advertising*, *45*(3), 334–345. https://doi.org/10.1080/00913367.2016.1172387

de Keyzer, F., van Noort, G., & Kruikemeier, S. (2022). Going too far? How consumers respond to personalized advertising from different sources. *Journal of Electronic Commerce Research*, *23*(3), 138–159. https://edepot.wur.nl/575826

Eisend, M., Niewiadomska, D., & van Noort, G. (2023, June 29–July 1). Personalization in marketing communication: A meta-analysis [Paper presentation]. The 21st International Conference on Research in Advertising 2023, Bordeaux, France.

Eslami, M., Rickman, A., Vaccaro, K., Aleyasen, A., Vuong, A., Karahalios, K., Hamilton, K., & Sandvig, C. (2015). "I always assumed that I wasn't really that close to [her]": Reasoning about invisible algorithms in news feeds. *Proceedings of the 33rd Annual ACM Conference on Human Factors in Computing Systems, South Korea*, 153–162. https://doi.org/10.1145/2702123.2702556

Frick, N. R., Wilms, K. L., Brachten, F., Hetjens, T., Stieglitz, S., & Ross, B. (2021). The perceived surveillance of conversations through smart devices. *Electronic Commerce Research and Applications*, *47*, Article 101046. https://doi.org/10.1016/j.elerap.2021.101046

Friestad, M., & Wright, P. (1994). The persuasion knowledge model: How people cope with persuasion attempts. *Journal of Consumer Research*, *21*(1), 1–31. https://doi.org/10.1086/209380

Gran, A.-B., Booth, P., & Bucher, T. (2021). To be or not to be algorithm aware: A question of a new digital divide? *Information, Communication & Society*, *24*(12), 1779–1796. https://doi.org/10.1080/1369118X.2020.1736124

Helberger, N., Dobber, T., & de Vreese, C. (2021). Towards unfair political practices law: Learning lessons from the regulation of unfair commercial practices for online political advertising. *JIPTEC*, *12*(3), 273–296. https://www.jipitec.eu/issues/jipitec-12-3-2021/5338

Helberger, N., Sax, M., Strycharz, J., & Micklitz, H.-W. (2022). Choice architectures in the digital economy: Towards a new understanding of digital vulnerability. *Journal of Consumer Policy*, *45*(2), 175–200. https://doi.org/10.1007/s10603-021-09500-5

Kitchin, R. (2017). Thinking critically about and researching algorithms. *Information, Communication & Society, 20*(1), 14–29. https://doi.org/10.1080/1369118X.2016.1154087

Kruikemeier, S., Sezgin, M., & Boerman, S. C. (2016). Political microtargeting: Relationship between personalized advertising on Facebook and voters' responses. *Cyberpsychology, Behavior, and Social Networking, 19*(6), 367–372. https://doi.org/10.1089/cyber.2015.0652

Limpf, N., & Voorveld, H. A. M. (2015). Mobile location-based advertising: How information privacy concerns influence consumers' attitude and acceptance. *Journal of Interactive Advertising, 15*(2), 111–123. https://doi.org/10.1080/15252019.2015.1064795

Maslowska, E., Smit, E. G., & van den Putte, B. (2016). It is all in the name: A study of consumers' responses to personalized communication. *Journal of Interactive Advertising, 16*(1), 74–85. https://doi.org/10.1080/15252019.2016.1161568

Peres, R., Schreier, M., Schweidel, D. A., & Sorescu, A. (2023). Blockchain meets marketing: Opportunities, threats, and avenues for future research. *International Journal of Research in Marketing, 40*(1), 1–11. https://doi.org/10.1016/j.ijresmar.2022.08.001

Richards, J. I., & Curran, C. M. (2002). Oracles on "advertising": Searching for a definition. *Journal of Advertising, 31*(2), 63–77. https://doi.org/10.1080/00913367.2002.10673667

Segijn, C. M., Strycharz, J., Riegelman, A., & Hennesy, C. (2021). A literature review of personalization transparency and control: Introducing the transparency–awareness–control framework. *Media and Communication, 9*(4), 120–133. https://doi.org/10.17645/mac.v9i4.4054

Segijn, C. M., & Voorveld, H. A. M. (2021). A first step in unraveling synced advertising effectiveness. *International Journal of Advertising, 40*(1), 124–143. https://doi.org/10.1080/02650487.2020.1778279

Segijn, C. M., Voorveld, H. A. M., & Vakeel, K. A. (2021). The role of ad sequence and privacy concerns in personalized advertising: An eye-tracking study into synced advertising effects. *Journal of Advertising, 50*(3), 320–329. https://doi.org/10.1080/00913367.2020.1870586

Senftleben, M., van Noort, G., & Smit, E. G. (2021). Towards AI transparency in brand-based communication—Evidence for better policymaking [Granted research project]. https://humane-ai.nl/news/new-seed-funded-projects-2021-2022/

Smit, E. G., van Noort, G., & Voorveld, H. A. M. (2014). Understanding online behavioural advertising: User knowledge, privacy concerns and online coping behaviour in Europe. *Computers in Human Behavior, 32*, 15–22. https://doi.org/10.1016/j.chb.2013.11.008

Starke, A. D., Asotic, E., Trattner, C., & van Loo, E. J. (2023). Examining the user evaluation of multi-list recommender interfaces in the context of healthy recipe choices. *ACM Transactions on Recommender Systems*. Advance online publication. https://doi.org/10.1145/3581930

Strycharz, J., & Segijn, C. M. (2022). The future of dataveillance in advertising theory and practice. *Journal of Advertising*, *51*(5), 574–591. https://doi.org/10.1080/00913367.2022.2109781

Strycharz, J., & Segijn, C. M. (2023). Consumer differences in chilling effects. In A. Vignolles & M. K. J. Waiguny (Eds.), *Advances in advertising research* (*Vol. XII*)*: Communicating, designing and consuming authenticity and narrative* (pp. 107–120). Springer Gabler. https://doi.org/10.1007/978-3-658-40429-1_8

Strycharz, J., van Noort, G., Smit, E., & Helberger, N. (2019). Protective behavior against personalized ads: Motivation to turn personalization off. *Cyberpsychology: Journal of Psychosocial Research on Cyberspace*, *13*(2), Article 1. https://doi.org/10.5817/CP2019-2-1

Susser, D. (2019). Invisible influence: Artificial intelligence and the ethics of adaptive choice architectures. *Proceedings of the 2019 AAAI/ACM Conference on AI, Ethics, and Society, USA*, 403–408. https://doi.org/10.1145/3306618.3314286

van der Goot, M. J. (2022). Source orientation, anthropomorphism, and social presence in human–chatbot communication: How to proceed with these concepts. *Publizistik*, *67*(4), 555–578. https://doi.org/10.1007/s11616-022-00760-w

van Noort, G., Himelboim, I., Martin, J., & Collinger, T. (2020). Introducing a model of automated brand-generated content in an era of computational advertising. *Journal of Advertising*, *49*(4), 411–427. https://doi.org/10.1080/00913367.2020.1795954

van Noort, G., & Meppelink, C. S. (2021, June 24–26). *How people cope with persuasion in the era of AI: A revised persuasion knowledge model* [Abstract presentation]. The 19th International Conference on Research in Advertising 2021.

Voorveld, H. A. M. (2019). Brand communication in social media: A research agenda. *Journal of Advertising*, *48*(1), 14–26. https://doi.org/10.1080/00913367.2019.1588808

Voorveld, H. A. M. (2022). Demystifying algorithmic persuasion on TikTok: A differential vulnerability perspective [NWO Social Science and Humanities (SSH) open competition proposal].

Voorveld, H. A. M., Meppelink, C. S., & Boerman, S. B. (2023). *Algorithmic persuasion in social media: Identifying user groups based on awareness, appropriateness and coping ability* [Manuscript submitted for publication]. Amsterdam School of Communication Research, University of Amsterdam.

Voorveld, H. A. M., van Noort, G., Muntinga, D. G., & Bronner, F. (2018). Engagement with social media and social media advertising: The differentiating role of

platform type. *Journal of Advertising, 47*(1), 38–54. https://doi.org/10.1080/00913367.2017.1405754
Walrave, M., Poels, K., Antheunis, M. L., van den Broeck, E., & van Noort, G. (2018). Like or dislike? Adolescents' responses to personalized social network site advertising. *Journal of Marketing Communications, 24*(6), 599–616. https://doi.org/10.1080/13527266.2016.1182938
Yun, J. T., Segijn, C. M., Pearson, S., Malthouse, E. C., Konstan, J. A., & Shankar, V. (2020). Challenges and future directions of computational advertising measurement systems. *Journal of Advertising, 49*(4), 446–458. https://doi.org/10.1080/00913367.2020.1795757
Zarouali, B, Boerman, S. C., Voorveld, H. A. M., & van Noort, G. (2022). The algorithmic persuasion framework in online communication: Conceptualization and a future research agenda. *Internet Research, 32*(4), 1076–1096. https://doi.org/10.1108/INTR-01-2021-0049
Zarouali, B., Dobber, T., de Pauw, G., & de Vreese, C. (2022). Using a personality-profiling algorithm to investigate political microtargeting: Assessing the persuasion effects of personality-tailored ads on social media. *Communication Research, 49*(8), 1066–1091. https://doi.org/10.1177/0093650220961965
Zhang, D., Hendriks, H., Boerman, S. C., Araujo, T., & Voorveld, H. A. M. (2023). A peak into individuals' perceptions of surveillance. In A. Vignolles & M. K. J. Waiguny (Eds.), *Advances in advertising research (Vol. XII): Communicating, designing and consuming authenticity and narrative* (pp. 163–178). Springer Gabler. https://doi.org/10.1007/978-3-658-40429-1_12
Zhang, D., van der Goot, M., Hendriks, H., Boerman, S. C., Araujo, T., & Voorveld, H. (2021, June 24–26). "They know everything": Folk theories, thoughts, and feelings about surveillance in media technologies [Paper presentation]. The 19th International Conference on Research in Advertising 2021.
Zhang, D., Voorveld, H. A. M., & Boerman, S. C. (2023). Privacy concerns matter, knowledge does not: Investigating effects of online behavioral advertising among Chinese and Dutch adults. *Journal of Current Issues & Research in Advertising.* Advance online publication. https://doi.org/10.1080/10641734.2023.2199812

About the authors

Guda van Noort is a Full Professor and holds the Chair in Persuasion & New Media Technologies at ASCoR. From 2015 to 2022 she was Director of the Foundation for Scientific Research on Commercial Communication (SWOCC). E-mail: G.vanNoort@uva.nl

Hilde Voorveld is an Associate Professor of Persuasive Communication & New Media Technologies at ASCoR and is the Programme Group Director of Persuasive Communication. She studies algorithmic persuasion and consumer responses to digital marketing communication. E-mail: H.A.M.Voorveld@uva.nl

Joanna Strycharz is an Assistant Professor of Persuasion & New Media Technologies at ASCoR. Combining traditional and computational methods, she studies algorithmic persuasion, its impact on consumers, their privacy, and consumer empowerment. E-mail: J.Strycharz@uva.nl

12. Human–Machine Communication

Jochen Peter, Theo Araujo, Carolin Ischen, Sonia Jawaid Shaikh, Margot J. van der Goot, and Caroline L. van Straten[1]

Abstract
Humans increasingly communicate with machines, such as chatbots and social robots. This has given rise to the field of human–machine communication. Since the second half of the 2010s, ASCoR researchers have embraced research on human–machine communication and studied communicating machines both from a theoretical and an empirical perspective. This chapter first highlights selected theoretical contributions of ASCoR researchers to human–machine communication and then outlines insights gained by empirical studies on conversational agents, social robots, as well as artificial intelligence more generally. The chapter subsequently identifies key empirical and theoretical challenges encountered by ASCoR researchers and concludes with a brief outline of two important issues for future research.

Keywords: technology, artificial intelligence, chatbots, human–robot interaction, automation

Introduction

Communication research has traditionally focused on interpersonal, computer-mediated, and mass communication. All these types of communication deal with communication between at least two humans (Guzman, 2018). In recent years, however, machines have become sophisticated enough to engage in meaningful communication with humans (Hepp, 2020), not least because of groundbreaking developments in machine learning and artificial intelligence. Responding to a broader "non-human turn" in intellectual endeavours in the

1 From the second author onwards, authors are listed alphabetically and contributed equally.

Peter, J. Araujo, T., Ischen, C., Shaikh, S. J., van der Goot, M. J., & van Straten, C. L. (2024). Human–machine communication. In T. Araujo & P. Neijens (Eds.), *Communication research into the digital society: Fundamental insights from the Amsterdam School of Communication Research* (pp. 205-220). Amsterdam University Press. https://doi.org/10.5117/9789048560592_ch12

social sciences and humanities (Grusin, 2015), which gravitate away from a primary focus on exclusively human phenomena as research subjects, communication scholars have, therefore, started to embrace human–machine communication (HMC) as an additional field of research (Guzman & Lewis, 2020). In this context, researchers have not only started to study the processes and effects of HMC but have also developed theoretical frameworks to do so.

Roughly since the second half of the 2010s, researchers in the Amsterdam School of Communication Research (ASCoR) have also turned toward studying HMC, dealing with various types of communicating machines, such as conversational agents, social robots, as well as artificial intelligence more broadly. This chapter highlights and summarises selected theoretical underpinnings of this research; describes its empirical contributions to our understanding of HMC; and identifies important empirical and theoretical challenges encountered by ASCoR researchers and HMC scholars more broadly. The chapter concludes with a brief outline of two important issues for future research. In its focus on ASCoR research, the chapter mainly features studies by ASCoR researchers and thus represents the HMC literature selectively.

Theoretical and conceptual underpinning

As a nascent field of research, HMC research needs to demarcate itself from other fields and, at the same time, develop theoretical frameworks that guide its research. Based on earlier work (Gunkel, 2012; Guzman, 2018) and focusing on social robots, Peter and Kühne (2018), for example, have argued that the study of communicating machines, such as social robots, distinguishes human communication research from other fields in communication research in at least three ways: First, studying communicating machines questions traditional notions of the communication partner. Whereas communication research typically assumes that communication partners are human, HMC research explicitly posits that machines can be our communication partners and, by extension, that humans and machines can engage in interactions that are meaningful to the humans involved. Second, when we study how humans communicate with machines, we question long-standing notions of the medium. The medium is no longer the object that transmits a human-made message (as it is typical of mass communication), but becomes itself a subject with which humans communicate (see also Zhao, 2006). Third, according to Peter and Kühne (2018), the study of communicating machines forces us to reconsider the limits of communication because the boundaries of HMC may no longer equal those of human–human communication. Machines may be, for example, superior to humans in how they process and retrieve information. This may inevitably change how we communicate—and what we define as the boundaries of communication. HMC may thus both challenge and extend how we think and conceptualise communication.

Several scholars have also called for the development of theoretical frameworks that focus on the specificities of human communication research (e.g., Fox & Gambino, 2021; Peter et al., 2019). An important step into this direction is the work on the "computers are social actors" (CASA) framework (Nass & Moon, 2000). This framework posits that if a "computer's cues are sufficient for humans to attribute humanness to it, we then mindlessly apply scripts from human–human interaction to the human–computer interaction" (Westerman et al., 2020, p. 396). Scholars, however, have criticised CASA for its anthropocentric bias (Gambino et al., 2020), and have extended its scope, for example, by specifying its propositions and generalising it to "media as social actors" (Lombard & Xu, 2021). In this context, van der Goot and Etzrodt (in press) have recently attempted to contextualise CASA theoretically. The authors contrast the original media equation paradigm, from which CASA developed (Reeves & Nass, 1996), with what they call, based on Turkle's (2007) work, the "media evocation paradigm." The media equation paradigm assumes that people equate media and machines with people if these media and machines provide social cues. As a result, people treat them mindlessly as if they were social actors (van der Goot & Etzrodt, in press). The media evocation paradigm, in contrast, proposes that media, and communicating machines, in particular, have an ambiguous ontological status somewhere between the animate and the inanimate and between thing and person. As a result, media and machines evoke mindful responses in which we rethink what a social actor is. In this view, machines are social actors to which/whom people respond socially (van der Goot & Etzrodt, in press). According to van der Goot and Etzrodt (in press), explicating the two paradigms more systematically may not only improve the rigor of theorising in HMC research, but also our understanding of social responses to machines.

Scholars have emphasised that research on HMC is often impaired by conceptual inconsistencies (e.g., Peter et al., 2019; van der Goot, 2022; van Straten, Peter, & Kühne, 2020). While some variation in the use of (the same) concepts may be understandable given the interdisciplinary roots of HMC research, it does reduce the comparability of studies and hampers cumulative research insights. Against this background, various researchers have started to clarify crucial concepts in HMC. Focusing on anthropomorphism, Kühne and Peter (2023), for example, have argued that previous conceptualisations of the construct tended to conflate antecedents and consequences with defining dimensions of anthropomorphism, which renders the concept's scope and content fuzzy. As an alternative and based on a theory of mind framework, they suggest that anthropomorphism can be seen as "attributing thinking, feeling, perceiving, desiring, and choosing" to communicating machines, such as social robots (Kühne & Peter, 2023, p. 42). Next to clarifying widely used concepts, ASCoR scholars have also turned their attention to somewhat neglected concepts. Van der Goot (2022), for instance, has emphasised that a broader understanding of how people respond to communicating machines requires a stronger focus on source orientation, that is, whether people orient themselves toward the manifest machine as the source of communication or toward latent sources, such as algorithms or the producers of the machines.

Researchers have observed that conceptualisation problems not only affect the constructs with which they try to understand HMC but also the very objects they aim to study. Addressing unclarities in the definition of a rapidly spreading category of communicating machines—intelligent assistants like Amazon's Alexa of Microsoft's Cortana—Shaikh (2023) has reviewed their evolution and existing definitions and concluded that such assistants can only be labelled intelligent when they are powered by artificial intelligence, interact with users, and assist them in various tasks. Similarly, Peter et al. (2019) distinguished social robots from advanced toys according to whether (1) they feature sensors, software control, interactivity, movement, embodiment, and energy; (2) the extent to which each of these features is present; and (3) their need for connectivity to function.

Overall, HMC scholars have proactively dealt with the theoretical and conceptual problems that the study of communicating machines entails. Although the field has progressed, researchers are confronted with at least two challenges. First, the technologies that drive communicating machines, notably machine learning and artificial intelligence, develop rapidly and often unpredictably, as do the ways in which people use and understand these technologies. What scholars study—the technology, the user, and the interplay between both—is thus subject to permanent change. Second, communicating machines themselves present us with hitherto unasked questions, for example, about their ontological and moral status. Accordingly, people may lack adequate ways of verbalising what they encounter in HMC and scholars may lack theories and concepts of how to investigate it best (van der Goot & Etzrodt, in press). How scholars study HMC thus requires theoretical and conceptual discussions that go beyond the field itself and require input from other disciplines, such as engineering and philosophy.

Empirical findings

As the number of communicating machines and their functionalities is increasing quickly, it remains difficult to single out the key lines of research in HMC. Two currently thriving research areas, however, centre on conversational agents and social robots. A third area deals with artificial intelligence more generally.

Conversational agents

A conversational agent can be defined as "any dialogue system that conducts natural language processing (NLP) and responds automatically using human language" (Griffing, 2022). When a conversational agent is disembodied, it can be text-based (i.e., chatbot) or voice-based (e.g., Siri). When a conversational agent is embodied, it can be graphically (i.e., on screens) or physically embodied (Griffing, 2022; see also Araujo, 2018). Researchers have predominantly dealt with disembodied conversational agents, typically in the form of chatbots (Følstad et al., 2021), focusing, for instance, on younger and older users (van der Goot & Pilgrim, 2020; Wald et al., 2023) as well as on political and commercial use contexts (Ischen et al., 2022; van der Goot et al., 2021; Zarouali et al., 2021).

While current research is diverse, three main lines permeate studies on conversational agents. A first line of research centres on the antecedents of the use of conversational agents. People's acceptance and use of a technology along with the question of what predicts it have been long-standing foci in technology research (Davis, 1989) and have also generated attention among researchers of conversational agents. In their survey-based research among families with small children, Wald et al. (2023), for example, found that the use of voice-based assistants is primarily hedonically motivated: Parents want to have fun when they use devices like Google Assistant with their children. In a teamwork setting, however, the use of voice-based assistants seems more functionally motivated. When under time pressure, teams use voice-based intelligent assistants more often than when they are not under time pressure, as Shaikh and Cruz (2022) showed in an experimental lab study.

The use of customer service chatbots is also mainly functional. Qualitative interview studies indicated that both older and younger people use such chatbots to receive answers to their queries in an easy and efficient manner (van der Goot & Pilgrim, 2020) and expect the interaction to be informative, fast, and friendly (van der Goot et al., 2021). The intended use of chatbots for sensitive purposes, such as mental health, finally, seems to be more complex. Based on an online survey among LGBTQIA+ individuals, Henkel et al. (2023) found that chatbot use is more likely when the chatbot is expected to perform well. In contrast, when people expect the chatbot to be difficult to use, they are less likely to use it. Characteristics such as an individual's willingness to self-disclose to a chatbot and the influence of significant others also increases the intention to use a chatbot (Henkel et al., 2023).

A second line of research on conversational agents investigates how they are experienced and perceived. Researchers often focus on concepts such as anthropomorphism, social presence, or interactivity, that are also theoretically important to other research on HMC. For instance, an online experiment comparing a chatbot and website unexpectedly found that the two did not differ in their perceived anthropomorphism and interactivity. The chatbot, however, was experienced as more enjoyable than the website (Ischen, Araujo, van Noort, et al., 2020). In another online experiment, a text-based chatbot was surprisingly seen as more human-like than a voice-based agent (Ischen et al., 2022). Not all text-based chatbots, however, perform equally: Araujo (2018) found in an experiment that a more human-like chatbot yielded higher levels of anthropomorphism than a machine-like chatbot. Interestingly, the effect on (mindless) anthropomorphism declined when the human-like chatbot was described as AI-powered. Social presence remained unaffected by the chatbot's human-like or machine-like character (Araujo, 2018). One explanation of the somewhat unexpected findings may be that users, at least of customer service chatbots, still seem to be frustrated by the limited functionality of chatbots (van der Goot et al., 2021; van der Goot & Pilgrim, 2020), which may affect the extent to which people experience them as human-like and socially present.

A third line of research on conversational agents deals with their effects. In a commercial context, a text-based chatbot was found to positively influence people's attitude toward a brand and their purchase intention because it was seen as more human-like than a voice-based agent (Ischen et al., 2022). Similarly, when people encountered a human-like chatbot rather than a machine-like chatbot, they disclosed more personal information and adhered more to its recommendations because they perceived it as more human-like and had fewer privacy concerns with it. The results, however, were the same when compared to a website as another experiment showed (Ischen, Araujo, Voorveld, et al., 2020). In a news context, by contrast, a news chatbot was more persuasive than a website article. In an experimental study, people accepted a counter-attitudinal article more easily when the chatbot provided the information. They also considered the information more credible because they (mindlessly) anthropomorphised the chatbot more than the website (Zarouali et al., 2021).

Social robots

Research on social robots has only advanced in the past 20 years or so. As a young and interdisciplinary field, dominant lines of research have yet to emerge, but some topics have attracted considerable research attention. Largely based on Peter (2022), we sketch research on the acceptance of, and relationship formation with, social robots. In our focus on ASCoR research, we limit ourselves to research done among children.

Similar to research on conversational agents, the question of when users accept social robots has become more central in research as social robots have become more widely accessible. Research that studied children's acceptance of social robots by surveying them before or after interactions with social robots has shown that children look primarily for hedonic gratifications in social robots. For children, social robots seem to be an object of play while social and informational gratifications are less important (de Jong et al., 2019; de Jong et al., 2021). However, prior to interacting with them, children hardly know what to expect from social robots (de Jong et al., 2019). Instead, children's intention to accept a robot is largely associated with more stable predictors, notably their general attitude toward robots, hedonic attitudes, as well as injunctive and descriptive norms about robot use (de Jong et al., 2022). Although children's acceptance of social robots is initially very high, it decays quickly, with a sharp drop after two weeks of acquiring a robot. Many children alternate between using and not using a robot, largely because robots face the competition of other toys and devices in children's (digital) environment (de Jong, 2022, ch. 5).

Social robots can affect various learning outcomes in children (e.g., prosocial behaviour) (Peter et al., 2021). The emergence of child–robot relationships is seen as crucial in this context and has been studied in a series of experiments. For example, similar to interpersonal relationships, child–robot relationships can be intensified when a robot asks a child questions. At the same time, a robot's self-disclosure reduced children's perceptions of the robot as being able to understand their feelings (van Straten et al., 2022). Likewise, when a robot specifically described itself, children perceived it as less similar to themselves (van Straten et al., 2021). Whereas self-disclosure and self-description are crucial to interpersonal relationships, neither affected the process of child–robot relationship formation itself. However, when children were informed about a robot's lack of human psychological abilities, either by a researcher or by a robot itself, child–robot relationship formation decreased (van Straten et al., 2023; van Straten, Peter, Kühne, et al., 2020), which supports calls for responsible robotics and ethics by design (see, e.g., de Pagter, 2023).

Artificial intelligence

Artificial intelligence (AI) largely drives the development of communicating machines and can be defined as "the study, understanding, and building of intelligent agents which can receive percepts from their environment(s), adapt, and act rationally, and make changes to their environments" (Shaikh, 2023, p. 784; see also de Vreese, 2024, this book). Given the far-reaching societal consequences of AI, ASCoR researchers have started to deal with people's attitudes toward AI. Based on a scenario-based survey experiment, Araujo et al. (2020), for instance, showed that people see risks in the use of AI in automated decision-making and doubt its fairness and usefulness. At the same time, people seem to evaluate automated decisions as equal to, or even better than, decisions by human experts. Using the same data, Helberger et al. (2020) found that automated decision-making is also considered as fairer than decisions taken by experts. To substantiate this judgement, people particularly rely on their assessment of the negative role of emotions in human decision-making and of the perceived objective (i.e., data-driven) character of automated decision-making, notably when it is correctly programmed (Helberger et al., 2020). Based on a content analysis of English-language media on the use of AI in the newsroom, Moran and Shaikh (2022) found opposing views among members of news organisations and journalists: News organisations see AI mostly as a cost-saving way of dealing with enormous amounts of information, which may enable journalist to do more substantive work. Journalists, in contrast, consider the products that AI generates to be of low quality and are worried about badly informed news consumers, while, simultaneously, they fear for their jobs.

Conclusion

All the research described in the previous sub-sections deals with advanced communicating machines or sophisticated technologies that rapidly change. ASCoR researchers have taken up this challenge and enriched our knowledge about antecedents of chatbot and robot acceptance and use; about people's perceptions of chatbots and social robots; and about the effects that chatbots and social robots have on issues as diverse as consumer behaviour, news reception, and relationship formation. We also start to recognise that people may have diverse, and even contradictory, attitudes toward technologies powered by AI.

At the same time, studying quickly changing communicating machines necessarily leads to "snapshot findings"—insights that are inevitably affected by the features and functionalities of a machine at a given point of time and, as a result, are often inconsistent and difficult to compare. While we understand, for example, that the acceptance and use of communicating machines depend, in their hedonic or functional orientation, on the context, we are still left with rather broad attitudes and norms as predictors of such acceptance and use. Similarly, although the importance of concepts such as social presence and notably anthropomorphism has been demonstrated repeatedly, it is still hard to pinpoint the exact features in communicating machines that increase or decrease them. Finally, compared to only a few years ago, we know now much more about the effects of conversational agents and social robots. However, many of these effects are based on the limited social abilities of these machines and we consequently lack insights into what communicating machines do with human sociality and human relationships more broadly. Overall, then, there is an urgent need for ongoing empirical research on HMC along with a keen eye on important technological developments.

Next steps: Development of a research infrastructure and an ethical framework

As research on HMC is still an emerging field, scholars have noted several problems in the field, ranging from methodological issues to theoretical challenges to cultural bias (e.g., Følstad et al., 2021; Peter et al., 2019). Partly in response to such current obstacles in the field, researchers have proposed avenues, or complete agendas, for future research on chatbots (Følstad et al., 2021), social robots (Peter & van Straten, in press), and AI and communication

(Guzman & Lewis, 2020; van der Goot & Etzrodt, in press). We agree with the suggestions that these researchers have made and support the turn toward more programmatic research on HMC. In this context, we believe that two basic aspects of HMC research deserve particular attention.

First, whenever communicating machines need to really act in studies (like in the studies on chatbots and social robots presented above), research is technologically demanding and resource intensive. We, therefore, need to establish research infrastructures that are easy to use and at the same time affordable. An example is the development of the Conversational Agent Research Toolkit (CART) (Araujo, 2020), with which researchers can rather easily create text-based chatbots for research. Research on social robots would also benefit from similar developments. Currently, researchers of social robots either rely on robots created in robotics institutes (e.g., Kaspar, University of Hertfordshire) or on commercially available, but often expensive robots (e.g., Nao, SoftBank). Findings, however, may depend heavily on the different types of robots used. A more homogenous infrastructure both in chatbot and social robot research may greatly advance the comparability of studies and the cumulative insights they generate.

Second, in the context of questions about the ontological status of communicating machines (van der Goot & Etzrodt, in press), researchers also face questions about the ethics of these machines and doing research with them. We, therefore, need an ethical framework that guides researchers, notably when we study machines or technologies that feature novel functionalities that may conflict with traditional ethical boundaries (e.g., advanced data collection or cognitive possibilities). Specifically, we need to focus on ethics by design and the challenge of how communicating machines can be designed such that they meet ethical principles (Følstad et al., 2021). Moreover, we need to consider more strongly potential privacy and security problems when using communicating machines in research, as well as when it is justified to deceive people about the features of communicating machines (Følstad et al., 2021; Peter & van Straten, in press; van der Goot, 2022).

As communicating machines become rapidly more advanced, communication researchers have started to chart this new field of research. ASCoR researchers have dealt with many of the technologies and developments and contributed to a better understanding of fundamental issues and pressing problems, both theoretically and empirically. With the progress in AI, the key driver of communicating machines, accelerating rather than slowing down, research on how humans and machines communicate with each other will become even more important in the future, not only within ASCoR, but also in communication research more broadly.

References

Araujo, T. (2018). Living up to the chatbot hype: The influence of anthropomorphic design cues and communicative agency framing on conversational agent and company perceptions. *Computers in Human Behavior, 85*, 183–189. https://doi.org/10.1016/j.chb.2018.03.051

Araujo, T. (2020). Conversational Agent Research Toolkit: An alternative for creating and managing chatbots for experimental research. *Computational Communication Research*, 2(1), 35–51. https://doi.org/10.5117/CCR2020.1.002.ARAU

Araujo, T., Helberger, N., Kruikemeier, S., & de Vreese, C. H. (2020). In AI we trust? Perceptions about automated decision-making by artificial intelligence. *AI & Society, 35*(3), 611–623. https://doi.org/10.1007/s00146-019-00931-w

Davis, F. D. (1989). Perceived usefulness, perceived ease of use, and user acceptance of information technology. *MIS Quarterly, 13*(3), 319–339. https://doi.org/10.2307/249008

de Jong, C. (2022). *Children and social robots: Towards a better understanding of their acceptance of a new technology* (Publication no. 9789493270718) [Doctoral dissertation, University of Amsterdam]. UvA-DARE. https://dare.uva.nl/search?field1=keyword;value1=de%20jong%20social%20robots;docsPerPage=1;startDoc=1

de Jong, C., Kühne, R., Peter, J., van Straten, C. L., & Barco, A. (2019). What do children want from a social robot? Toward gratifications measures for child–robot interaction. *Proceedings of the 2019 28th IEEE International Conference on Robot and Human Interactive Communication (RO-MAN), India*, 1–8. https://doi.org/10.1109/RO-MAN46459.2019.8956319

de Jong, C., Peter, J., Kühne, R., & Barco, A. (2022). Children's intention to adopt social robots: A model of its distal and proximal predictors. *International Journal of Social Robotics, 14*(4), 875–891. https://doi.org/10.1007/s12369-021-00835-0

de Jong, C., Peter, J., Kühne, R., van Straten, C., & Barco, A. (2021). Exploring children's beliefs for adoption or rejection of domestic social robots. *Proceedings of the 2021 30th IEEE International Conference on Robot & Human Interactive Communication (RO-MAN)*, 598–603. https://doi.org/10.1109/RO-MAN50785.2021.9515438

de Pagter, J. (2023). From EU robotics and AI governance to HRI research: Implementing the ethics narrative. *International Journal of Social Robotics*. Advance online publication. https://doi.org/10.1007/s12369-023-00982-6

de Vreese, C. H. (2024) How artificial intelligence is changing ASCoR's research. In T. Araujo & P. Neijens (Eds.), *Communication research into the digital society: Fundamental insights from the Amsterdam School of Communication Research* (pp. 221–231). Amsterdam University Press. https://doi.org/10.5117/9789048560592_ch13

Følstad, A., Araujo, T., Law, E. L.-C., Brandtzaeg, P. B., Papadopoulos, S., Reis, L., Baez, M., Laban, G., McAllister, P., Ischen, C., Wald, R., Catania, F., Meyer von Wolff, R., Hobert, S., & Luger, E. (2021). Future directions for chatbot research: An interdisciplinary research agenda. *Computing, 103*(12), 2915–2942. https://doi.org/10.1007/s00607-021-01016-7

Fox, J., & Gambino, A. (2021). Relationship development with humanoid social robots: Applying interpersonal theories to human/robot interaction. *Cyberpsychology, Behavior, and Social Networking, 25*(5), 294–299. https://doi.org/10.1089/cyber.2020.0181

Gambino, A., Fox, J., & Ratan, R. (2020). Building a stronger CASA: Extending the computers are social actors paradigm. *Human–Machine Communication, 1*, 71–86. https://doi.org/10.30658/hmc.1.5

Griffing, E. (2022, December 2). *What is a conversational agent?* Dashbot. https://www.dashbot.io/blog/conversational-agent

Grusin, R. (Ed.). (2015). *The nonhuman turn.* University of Minnesota Press.

Gunkel, D. J. (2012). Communication and artificial intelligence: Opportunities and challenges for the 21st century. *Communication + 1, 1*(1), Article 1. https://doi.org/10.7275/R5QJ7F7R

Guzman, A. L. (2018). What is human–machine communication, anyway? In A. L. Guzman (Ed.), *Human–machine communication: Rethinking communication, technology, and ourselves* (pp. 1–28). Peter Lang.

Guzman, A. L., & Lewis, S. C. (2020). Artificial intelligence and communication: A human–machine communication research agenda. *New Media & Society, 22*(1), 70–86. https://doi.org/10.1177/1461444819858691

Helberger, N., Araujo, T., & de Vreese, C. H. (2020). Who is the fairest of them all? Public attitudes and expectations regarding automated decision-making. *Computer Law & Security Review, 39*, Article 105456. https://doi.org/10.1016/j.clsr.2020.105456

Henkel, T., Linn, A. J., & van der Goot, M. J. (2023). Understanding the intention to use mental health chatbots among LGBTQIA+ individuals: Testing and extending the UTAUT. In A. Følstad, T. Araujo, S. Papadopoulos, E. L.-C. Law, E. Luger, M. Goodwin, & P. B. Brandtzaeg (Eds.), *Chatbot research and design* (pp. 83–100). Springer International Publishing. https://doi.org/10.1007/978-3-031-25581-6_6

Hepp, A. (2020). Artificial companions, social bots and work bots: Communicative robots as research objects of media and communication studies. *Media, Culture & Society, 42*(7–8), 1410–1426. https://doi.org/10.1177/0163443720916412

Ischen, C., Araujo, T. B., Voorveld, H. A. M., van Noort, G., & Smit, E. G. (2022). Is voice really persuasive? The influence of modality in virtual assistant interactions and two alternative explanations. *Internet Research, 32*(7), 402–425. https://doi.org/10.1108/INTR-03-2022-0160

Ischen, C., Araujo, T., van Noort, G., Voorveld, H., & Smit, E. (2020). "I am here to assist you today": The role of entity, interactivity and experiential perceptions in chatbot persuasion. *Journal of Broadcasting & Electronic Media, 64*(4), 615–639. https://doi.org/10.1080/08838151.2020.1834297

Ischen, C., Araujo, T., Voorveld, H., van Noort, G., & Smit, E. (2020). Privacy concerns in chatbot interactions. In A. Følstad, T. Araujo, S. Papadopoulos, E. L.-C. Law, O.-C. Granmo, E. Luger, & P. B. Brandtzaeg (Eds.), *Chatbot research and design* (pp. 34–48). Springer International Publishing. https://doi.org/10.1007/978-3-030-39540-7_3

Kühne, R., & Peter, J. (2023). Anthropomorphism in human–robot interactions: A multidimensional conceptualization. *Communication Theory, 33*(1), 42–52. https://doi.org/10.1093/ct/qtac020

Lombard, M., & Xu, K. (2021). Social responses to media technologies in the 21st century: The media are social actors paradigm. *Human–Machine Communication, 2*, 29–55. https://doi.org/10.30658/hmc.2.2

Moran, R. E., & Shaikh, S. J. (2022). Robots in the news and newsrooms: Unpacking meta-journalistic discourse on the use of artificial intelligence in journalism. *Digital Journalism, 10*(10), 1756–1774. https://doi.org/10.1080/21670811.2022.2085129

Nass, C., & Moon, Y. (2000). Machines and mindlessness: Social responses to computers. *Journal of Social Issues, 56*(1), 81–103. https://doi.org/10.1111/0022-4537.00153

Peter, J. (2022). Social robots and children. In D. Lemish (Ed.), *The Routledge international handbook of children, adolescents, and media* (2nd ed., pp. 153–161). Routledge.

Peter, J., & Kühne, R. (2018). The new frontier in communication research: Why we should study social robots. *Media and Communication, 6*(3), 73–76. https://doi.org/10.17645/mac.v6i3.1596

Peter, J., Kühne, R., & Barco, A. (2021). Can social robots affect children's prosocial behavior? An experimental study on prosocial robot models. *Computers in Human Behavior, 120*, Article 106712. https://doi.org/10.1016/j.chb.2021.106712

Peter, J., Kühne, R., Barco, A., de Jong, C., & van Straten, C. L. (2019). Asking today the crucial questions of tomorrow: Social robots and the Internet of Toys. In G. Mascheroni & D. Holloway (Eds.), *The internet of Toys: Practices, affordances and the political economy of children's smart play* (pp. 25–46). Palgrave Macmillan. https://doi.org/10.1007/978-3-030-10898-4_2

Peter, J., & van Straten, C. L. (In press). Social robots and children: A field in development. In L. Fortunati & A. Edwards (Eds.), *De Gruyter handbook of robots in society and culture*. De Gruyter Mouton.

Reeves, B., & Nass, C. (1996). *The media equation: How people treat computers, television, and new media like real people and places*. Cambridge University Press.

Shaikh, S. J. (2023). Artificially intelligent, interactive, and assistive machines: A definitional framework for intelligent assistants. *International Journal of*

Human–Computer Interaction, *39*(4), 776–789. https://doi.org/10.1080/10447318.2022.2049133

Shaikh, S. J., & Cruz, I. F. (2022). AI in human teams: Effects on technology use, members' interactions, and creative performance under time scarcity. *AI & Society*. Advance online publication. https://doi.org/10.1007/s00146-021-01335-5

Turkle, S. (Ed.). (2007). *Evocative objects: Things we think with*. MIT Press.

van der Goot, M. J. (2022). Source orientation, anthropomorphism, and social presence in human–chatbot communication: How to proceed with these concepts. *Publizistik*, *67*(4), 555–578. https://doi.org/10.1007/s11616-022-00760-w

van der Goot, M. J., & Etzrodt, K. (In press). Disentangling two fundamental paradigms in human–machine communication research: Media equation and Media evocation. *Human–Machine Communication*.

van der Goot, M. J., Hafkamp, L., & Dankfort, Z. (2021). Customer service chatbots: A qualitative interview study into the communication journey of customers. In A. Følstad, T. Araujo, S. Papadopoulos, E. L.-C. Law, E. Luger, M. Goodwin, & P. B. Brandtzaeg (Eds.), *Chatbot research and design* (pp. 190–204). Springer International Publishing. https://doi.org/10.1007/978-3-030-68288-0_13

van der Goot, M. J., & Pilgrim, T. (2020). Exploring age differences in motivations for and acceptance of chatbot communication in a customer service context. In A. Følstad, T. Araujo, S. Papadopoulos, E. L.-C. Law, O.-C. Granmo, E. Luger, & P. B. Brandtzaeg (Eds.), *Chatbot research and design* (pp. 173–186). Springer International Publishing. https://doi.org/10.1007/978-3-030-39540-7_12

van Straten, C. L., Peter, J., & Kühne, R. (2020). Child–robot relationship formation: A narrative review of empirical research. *International Journal of Social Robotics*, *12*(2), 325–344. https://doi.org/10.1007/s12369-019-00569-0

van Straten, C. L., Peter, J., & Kühne, R. (2023). Transparent robots: How children perceive and relate to a social robot that acknowledges its lack of human psychological capacities and machine status. *International Journal of Human–Computer Studies*, *177*, Article 103063. https://doi.org/10.1016/j.ijhcs.2023.103063

van Straten, C. L., Peter, J., Kühne, R., & Barco, A. (2020). Transparency about a robot's lack of human psychological capacities: Effects on child–robot perception and relationship formation. *ACM Transactions on Human–Robot Interaction*, *9*(2), Article 11. https://doi.org/10.1145/3365668

van Straten, C. L., Peter, J., Kühne, R., & Barco, A. (2021). The wizard and I: How transparent teleoperation and self-description (do not) affect children's robot perceptions and child–robot relationship formation. *AI & Society*, *37*(1), 383–399. https://doi.org/10.1007/s00146-021-01202-3

van Straten, C. L., Peter, J., Kühne, R., & Barco, A. (2022). On sharing and caring: Investigating the effects of a robot's self-disclosure and question-asking on children's robot perceptions and child–robot relationship formation. *Computers in Human Behavior, 129*, Article 107135. https://doi.org/10.1016/j.chb.2021.107135

Wald, R., Piotrowski, J. T., Araujo, T., & van Oosten, J. M. F. (2023). Virtual assistants in the family home: Understanding parents' motivations to use virtual assistants with their child(dren). *Computers in Human Behavior, 139*, Article 107526. https://doi.org/10.1016/j.chb.2022.107526

Westerman, D., Edwards, A. P., Edwards, C., Luo, Z., & Spence, P. R. (2020). I-It, I-Thou, I-Robot: The perceived humanness of AI in human–machine communication. *Communication Studies, 71*(3), 393–408. https://doi.org/10.1080/10510974.2020.1749683

Zarouali, B., Makhortykh, M., Bastian, M., & Araujo, T. (2021). Overcoming polarization with chatbot news? Investigating the impact of news content containing opposing views on agreement and credibility. *European Journal of Communication, 36*(1), 53–68. https://doi.org/10.1177/0267323120940908

Zhao, S. (2006). Humanoid social robots as a medium of communication. *New Media & Society, 8*(3), 401–419. https://doi.org/10.1177/1461444806061951

About the authors

Jochen Peter is a Full Professor and holds the Chair in Media Entertainment at ASCoR. He is Programme Group Director of Youth & Media Entertainment at ASCoR. He was Scientific Director of ASCoR from 2013 to 2017. His research centres on emerging technologies and the psycho-social development of young people. E-mail: J.Peter@uva.nl

Theo Araujo is a Full Professor of Media, Organisations and Society, and Scientific Director of ASCoR as of 2022. His research investigates the dynamic interplay between media and organisations, and what it means for society, with a special focus on trust and technology. His research has also a methodological focus on computational communication science and the implementation of large-scale data collection and analysis for communication research. E-mail: T.B.Araujo@uva.nl

Carolin Ischen is an Assistant Professor of Persuasive Communication at ASCoR. Her research focuses on emerging media technologies such as conversational agents. She is interested in how the characteristics of non-human communication partners influence human interactions with them, as well as the persuasive consequences of these interactions. E-mail: C.Ischen@uva.nl

Sonia Jawaid Shaikh is an Assistant Professor of Corporate Communication at ASCoR. Her research focuses on intelligent technologies, human decision-making, and the news media. E-mail: S.J.Shaikh@uva.nl

Margot van der Goot is a Senior Assistant Professor of Persuasion & New Media Technologies at ASCoR. She studies user perceptions of interactions with conversational agents, herein focusing on concepts such as source orientation, anthropomorphism, social presence, media equation, and media evocation. E-mail: M.J.vanderGoot@uva.nl

Caroline L. van Straten is an Assistant Professor of Youth & Media Entertainment at ASCoR. Her research investigates the emergence of social relationships between children and robots as well as communication between humans and machines more broadly. E-mail: C.L.vanStraten@uva.nl

13. How Artificial Intelligence Is Changing ASCoR's Research

Claes H. de Vreese

Abstract
Artifical intelligence is becoming an important theme in ASCoR research. ASCoR researchers are studying public perceptions of AI, AI in recommender systems, AI and journalism, AI in chatbots, and AI in relation to mis- and disinformation. Meanwhile, AI is also being leveraged methodologically in research. Research on AI-related topics are often carried out in larger interdisciplinary projects with colleagues in, for example, the University of Amsterdam's Institute for Information Law (IViR).

Keywords: recommender systems, large language models, journalism, media, trust

Introduction: The arr-AI-val of AI research

The impact of artificial intelligence (AI)[1] as an object of study and as methods deployed has become notable in the communication science field in recent years. Researchers across ASCoR are engaged in understanding these developments, both within and in conjunction with adjacent fields like law and media studies. AI in communication science has traditionally not been a central feature. In fact, despite the scientific attention to AI since the Second World War, AI research was scattered and it was not until

1 "Artificial intelligence" is defined broadly by the European Union in its white paper and AI Act as a collection of technologies that combine data, algorithms, and computing power. It refers to a system's ability to interpret data, to learn from such data, and to use those learnings to achieve specific goals and tasks through flexible adaptation (https://commission.europa.eu/system/files/2020-02/commission-white-paper-artificial-intelligence-feb2020_en.pdf).

de Vreese, C. H. (2024). How artificial intelligence is changing ASCoR's research. In T. Araujo & P. Neijens (Eds.), *Communication research into the digital society: Fundamental insights from the Amsterdam School of Communication Research* (pp. 221-231). Amsterdam University Press. https://doi.org/10.5117/9789048560592_ch13

the 1980s that the AI field had the features of a scientific discipline. As van den Besselaar and Leydesdorff (1996, p. 415), both affiliated with ASCoR for parts of their careers, concluded in an extensive bibliometric analysis, “AI has emerged as a set of journals with the characteristics of a discipline only since 1988.”

In the communication science field research is now emerging. Only a bit more than a decade ago, Gunkel (2012, p. 1) called for attention to be paid specifically to AI and argued “that communication studies need to rework its basic framework in order to address and respond to the unique technological challenges and opportunities of the 21st century.” About a decade later, there was still only an emerging body of work, which led Nah et al. (2020, p. 369) to conclude that the intersection AI and communication “remains ripe for exploration.”

Work in related areas such as human–machine communication (HMC), computer-mediated communication (CMC) meanwhile was sprawling and also influenced the research at ASCoR (see chapters by Neijens & Valkenburg, 2024, and by Peter et al., 2024, in this volume, and earlier work by, e.g., Tanis & Postmes, 2003). However, research taking a starting point in AI techniques such as machine learning, natural language processing, large language models (LLMs), etc., and the ways in which messages are generated, modified, or augmented as part of self-learning systems has only recently emerged on the agenda.

The potential impact of current developments is significant. Sundar (2020, p. 74) observes that developments in personalisation, algorithms, and machine learning “vastly enhances the ease and convenience of our media and communication experiences, but they have also raised significant concerns about privacy, transparency of technologies and human control over their operations.” These concerns are inherently interdisciplinary questions. This need for interdiscplinarity in addressing the concerns have also shaped the uptake of AI-related questions at ASCoR and the University of Amsterdam more widely, which will be discussed below.

On a more general level, Guzman and Lewis (2020) offer a clarifying take on the ways in which AI technologies affect the field of communication by distinguishing its functional, relational, and metaphysical aspects: They identify

> (1) the *functional* dimensions through which people make sense of these devices and applications as communicators, (2) the *relational* dynamics through which people associate with these technologies and, in turn, relate to themselves and others, and (3) the *metaphysical* implications

called up by blurring ontological boundaries surrounding what constitutes human, machine, and communication.

Traces of all three aspects can be found in "ASCoR's AI research."

Emerging AI trends at ASCoR

ASCoR and wider UvA-based research focuses, without being comprehensive, on at least the following five areas: (1) public perceptions of AI, (2) AI in journalism, (3) advanced recommender systems, (4) chatbots, and (5) AI-driven mis- and disinformation research. These cover both the functional aspect (how citizens and users make sense of AI), the relational aspect (how citizens and user use and relate with these technologies) and the metaphysical aspect of what constitutes the human and the AI aspects of communication. In addition, AI developments offer new additions to our methodological toolkit.

Public perceptions of AI: With the increase in data-driven, self-learning systems and automated decision-making, it is imperative to study the (changing) nature of how different (groups of) citizens perceive AI decision-making. Araujo et al. (2020) dove into this with a comprehensive study of perceptions of automated decision-making by AI and whether such perceptions differ across media, (public) health, and legal contexts. They examined users' perceptions of automated decision-making (ADM) by algorithms. They found that respondents are divided on the potential usefulness or fairness of automated decision-making processes and are concerned about potential risks. However, when comparing respondents' perceptions of fairness, usefulness, and risk for specific decisions within media, public health, and justice, ADM was for the most part seen as equal and at times rated better than human experts. This underscores the complexity of public opinion in this space, with many citizens holding mixed perceptions, but also a base level of trust in AI decisions. This study was carried out in 2018 and begs for a replication, as the AI discussion has become much more central in public discourse.

Research in the Digital Communication Methods Lab (digicomlab), funded initially by the Faculty of Social and Behavioural Sciences, as part of strategic investments (a so-called research priority area, or RPA) and now an integral part of ASCoR, also pivoted towards AI research. Work on digital competencies has innovated to include an "across-the-lifespan" perspective (in the Ministry of Interior Affairs-funded project DigiQ, with

Jessica Taylor Piotrowski as principal investigator and Dian de Vries as postdoc). Based on a survey with a nationally representative sample as part of the DigiQ project, ASCoR PhD student Chenyue Wang (Wang et al., 2023) identified five user groups based on users' AI-related knowledge, skills, and attitudes: the average users, the expert advocates, the expert sceptics, the unskilled sceptics, and the neutral unskilled. This finding resonates with existing research on the digital divide and provides evidence for an emerging AI divide among citizens. Especially elderly and less-educated users are significantly more likely to belong to vulnerable, lower skilled, less knowledgeable groups than the average user group, underscoring the need for interventions in this area.

AI in journalism. The long-standing tradition for research on (changes in) journalism in ASCoR also extends to questions around AI in journalism. In the AI, Media & Democracy (AIMD) Lab—funded by the Dutch Research Council (now) and directed by Natali Helberger (University of Amsterdam's Institute for Information Law, IViR) and Claes H. de Vreese—the role of AI in the production, distribution, and use of news is central. Exciting work is carried out with media organisations like DPG Media and the BBC. Postdoc in the lab and ASCoR member Hannes Cools conducted a study of innovation labs (internal development units) at the *Washington Post* and found that while engineering teams at the newspaper exert great efforts to make some algorithmic systems transparent and explain their functions to the public, less information is being shared inside the newsroom. This lack of internal transparency is a potential pitfall as it could lead to mistakes in the production of the news and the reporting process in general. This work points to the role of journalistic organisation in both covering, explaining, implementing, and discussing AI tools.

In an analysis of the use of AI in newsrooms, Moran and Shaikh (2022) identified internal organisational tensions. On the one hand, members of news management see AI as comprising tools that enable cost savings and efficiency, thus creating space for more meaningful journalistic work. Journalists themselves, on the other hand, see using AI tools as cumbersome, cause them to deliver sub-optimal work, and can lead to misinformation. It should be noted that these conclusions were drawn before the 2022 roll-out of generative AI models (like ChatGPT) on a large scale, which might change or exacerbate some of these patterns. Work by colleagues at IViR, with the participation of former ASCoR colleague Judith Möller, showed that a normative perspective on journalistic AI, which explicates values and tests AI usage on theses values, is needed. Such a perspective should be "grounded in a broader normative framework and a thorough understanding of the dynamics and complexities of journalistic AI at the level of people, newsrooms and media markets" (Helberger et al., 2022, p. 1605). The AI and journalism line of research is likely to expand in the coming years, as generative AI tools become available for both text, visuals, and sound content in journalism.

AI in recommender systems. In the past five years, a number of projects have focused on news recommender systems (NRS)—automated recommendations of news stories to users—from different angles. In the best situations, NRS can help guide citizens in an information rich ecosystem, with potential democratic benefits of this guidance (Mitova et al., 2023). While not situated as "AI research" as such, several of these studies make reference to and bring in AI techniques. News recommenders may indeed be AI driven. The ethical implications of when to use what systems have been laid out by Helberger et al. (2020), who also outline how AI tools can impact on fundamental rights, such as universal human rights. As the AI underpinnings become more cental, important normative questions are foregrounded: As Vrijenhoek et al. (2021, p. 173) (from the AIMD lab) with other ASCoR and IvIR colleagues put it: "Current approaches to evaluating recommender systems are often focused on measuring an increase in user clicks and short-term engagement, rather than measuring the user's longer term interest in diverse and important information." According to the authors, what is needed are new metrics for evaluating NRS and their implications.

AI chatbots. Conversation agents (CAs) (like Siri and Alexa; see the chapter by Peter et al., 2024) have important AI features such as self-learning systems for news provision. In an experiment in the PhD project of Valeria Resendez Gomez it was found that users relying on conversational agents for news have less agreement about the important issues for a country. These results imply that individuals mainly informing themselves through CAs might have a more diverse view on the relevant issues in the country than individuals using news websites for news consumption. Conversely, news website users in comparison to CAs have a higher probability of agreeing on the important events that are relevant for a country (Resendez et al., 2023). This dovetails with the idea that news organisations are mainly driven by journalistic values where publishers select news stories with diverse subjects and perspectives while promoting a shared public sphere.

AI-driven mis- and disinformation research. ASCoR has a long-standing tradition in trying to understand the persuasive influences of information, ranging from health information and advertising to news and journalism. With the advent of AI and especially generative AI techniques, there is a lot of concerns about a proliferation of mis- and disinformation. These new AI developments are also present in ASCoR research. ASCoR participates both in the Dutch–Flemish BENEDMO hub on disinformation and in the EU-funded EDMO (European Digital Media Observatory, 2023), where Claes H. de Vreese is on the executive board. Dobber et al. (2021), as part of the interdisciplinary initiative with the IViR, conducted one of the first studies of the impact of deepfakes (manipulated, AI-generated visual content). Participants who saw this deepfake, in which a Christian politician seemed to make a bad joke about Jesus Christ's suffering, held more negative attitudes toward the politician than the people who viewed a neutral control video. So the study established for the first time that deepfakes indeed have the potential to cause harm. This line of research is expanded and Hameleers et al. (2022), who found that deepfakes are, contrary to expectations and public concerns, not credible when compared to text or image-only disinformation. The affordances of AI do not add persuasive power to disinformation.

In sum, these five lines of research showcase the breadth of ongoing, ASCoR-based "AI research." At the time of writing (July 2023), much attention is pivoting towards understanding the affordances, their use, and the implications of LLMs (such as ChatGPT). As de Vreese and Votta (2023) have suggested these techniques give rise to a range of new empirical and normative reseach questions. These include, for example, questions around ethics and practices of political campaigning and microtargeting: AI tools will offer further opportunities for matching voters with tailored content, for optimising ad pricing, and for autogenerating content from generative AI applications. Other questions evolve around authenticity and what is "real" will explode with the rapid roll-out of accessible generative AI tools, for text, audio, and images. These are just some of the questions being addressed.

AI and methods

The study of AI and communication is a two-edged sword when it comes to methods. On the one hand, "classic" communication science and social science methods are being employed to study the new phenomena. This includes survey-based work (tapping attitudes, skills, etc.), experiments, interviews, and quasi-ethnographic news organisation work. Hannes Cools, for example, used a digital ethnography approach in studying the *Washington Post*, which included being present in key meeting and also access to, e.g., internal online discussion channels (like Slack), in addition to interviews (Cools et al., 2022).

The research has also increasingly considered AI as part of the methodological toolkit. Work in ASCoR on content analysis draws on neural topic modelling (BERTopic) in combination with network-analytic approaches to understand information flows. This approach allows for dynamically understanding how topics change over time (Simon et al., 2022). Developments in LLMs lead to using machine learning for very abstract and latent concepts as well, rather than for manifest indicators. A model can now distinguish between factual or opiniated genres, or formal and informal ones, do so on a large scale, and for different kinds of texts: video subtitles, podcasts, newspaper articles, blogposts, etc. (Lin et al., 2023; Trilling et al., 2024, in this volume).

On the visual side of things, research can now use deep learning to create hyper-realistic political videos which represent a novelty in the field of disinformation. In addition, de Vreese and Votta (2023) offered examples of new techniques that are available. We can study how AI-generated or partially AI-assisted news articles differ from human-written ones in terms of content, tone, and biases; we can explore how AI-driven recommendation systems affect the diversity and quality of political content users are exposed to, we can use LLM to measure the latent ideology of groups and individuals, and we can use LLM rather than training a custom machine learning model with code. These are just some examples of likely AI application in research methods.

Contributions to theory and practice

So far, the most recent "AI turn in communication research" is still rather nascent. There is work done in articulating the distinct communication science perspective and theorising on current developments. This reflection is needed to not merely respond to recent technological roll-outs but actually integrate them more in existing theory and concepts. ASCoR research, traditionally, is likely to gravitate towards empirical assessments, and put user perspectives central in the research. These theoretical reflections are likely to develop further in the coming years. This work may benefit from Guzman and Lewis (2020), who provide starting points for articulating when and how communication and AI is different from other technologies. Their three dimensions—functional, relational, and metaphysical, as discussed at the outset—serve as a useful anchor.

Several of the (partial) ASCoR-based AI research projects have a direct bearing on practice and regulation. The work on AI and journalism, for example, is often done with external stakeholders and media partners. There are huge questions in this sector currently, and they are exacerbated by generative AI technologies. The research acknowledges that there is a delicate, socio-technical dynamic at play. Solutions are likely to come from partnerships with media companies. This provides access to organisations, data, and real-world challenges, and they provide a lens through which to offer results that can be implemented in work routines, or inform strategic discussion and management decisions.

Next steps

The proliferation of AI tools in communication, journalism, education, training, politics, and advertising is already affecting our field. There are many (both old and new) substantive questions emerging. In answering these, communication scholars are well advised to collaborate both with, for example, experts on the ethical implications of using AI chatbots in (political) campaigns or debates and with legal and governance scholars who can situate the discussion in broader developments around the regulation of AI, such as the European Union's AI Act. Some contours of ASCoR-based communication research are emerging. AI and communication research will:

- Be *interdisciplinary* in nature when answering substantive questions, with, for example, links to political science (around political campaigns) and other social sciences, to law and regulation (and ethical, legal, and societal perspectives), and to AI groups interested in foundational models, also called general purpose AI models.
- Foreground work on *authenticity* in the context of mis- and disinformation. AI will be leveraged to detect inauthentic text, visuals, and sound content, but the technologies to create these are likely to develop faster, in combination with advanced amplification techniques and advances in recommender systems.
- Foreground questions about *trust*, both vis-à-vis journalistic content, but also more broadly in the information system. Research will probe how discussions about eroding trust might themselves have harmful and unintended consequences, reaching beyond, for example, the impact of AI-generated deepfakes.
- Affect *research* on augmented and virtual realities and metaverse developments by asking questions about how the technology can be

used to affect and impact citizens and consumers towards healthier and more sustainable behaviour.

- Be affected by the possibilities to use AI techniques to further *study* communication practices and patterns, using large-scale computational methods to integrate larger and multiple data sources.
- Centre questions about transparent and *explainable* AI (XAI) systems that provide citizens and users explanations for AI-driven decisions. Questions about inequities and different vulnerabilities will be central in this space and AI discussions will merge with current work on media and digital literacy.
- Push *institutional* boundaries further. Much research will have a centre in ASCoR, but collaboration across fields and disciplines will also mean that university-based, cross-disciplinary initiatives will be central hubs for such research.

References

Araujo, T., Helberger, N., Kruikemeier, S., & de Vreese, C. H. (2020). In AI we trust? Perceptions about automated decision-making by artificial intelligence. *AI & Society, 35*(3), 611–623. https://doi.org/10.1007/s00146-019-00931-w

Cools, H., van Gorp, B., & Opgenhaffen, M. (2022). New organizations, different journalistic roles, and innovative projects: How second-generation newsroom innovation labs are changing the news ecosystem. *Journalism Practice.* Advance online publication. https://doi.org/10.1080/17512786.2022.2097940

de Vreese, C., & Votta, F. (2023). AI and political communication. *Political Communication Report,* (27). https://doi.org/10.17169/refubium-39047

Dobber, T., Metoui, N., Trilling, D., Helberger, N., & de Vreese, C. (2021). Do (micro-targeted) deepfakes have real effects on political attitudes? *The International Journal of Press/Politics, 26*(1), 69–91. https://doi.org/10.1177/1940161220944364

European Digital Media Observatory. (2023, June 9). *Disinformation narratives about the war in Ukraine.* https://edmo.eu/war-in-ukraine-the-fact-checked-disinformation-detected-in-the-eu/

Gunkel, D. J. (2012). Communication and artificial intelligence: Opportunities and challenges for the 21st Century. *Communication + 1, 1*(1), Article 1. https://doi.org/10.7275/R5QJ7F7R

Guzman, A. L., & Lewis, S. C. (2020). Artificial intelligence and communication: A human–machine communication research agenda. *New Media & Society, 22*(1), 70–86. https://doi.org/10.1177/1461444819858691

Hameleers, M., van der Meer, T. G., & Dobber, T. (2022). You won't believe what they just said! The effects of political deepfakes embedded as vox populi on social media. *Social Media + Society*, *8*(3), Article 20563051221116346. https://doi.org/10.1177/20563051221116346

Helberger, N., van Drunen, M., Moeller, J., Vrijenhoek, S., & Eskens, S. (2022). Towards a normative perspective on journalistic AI: Embracing the messy reality of normative ideals. *Digital Journalism*, *10*(10), 1605–1626. https://www.tandfonline.com/doi/full/10.1080/21670811.2022.2152195

Lin, Z., Welbers, K., Vermeer, S., & Trilling, D. (2023). Beyond discrete genres: Mapping news items onto a multidimensional framework of genre cues. *Proceedings of the 17th International AAAI Conference on Web and Social Media, Cyprus*, *17*, 542–553. https://doi.org/10.48550/arxiv.2212.04185

Mitova, E., Blassnig, S., Strikovic, E., Urman, A., Hannak, A., de Vreese, C. H., & Esser, F. (2023). News recommender systems: A programmatic research review. *Annals of the International Communication Association*, *47*(1), 84–113. https://doi.org/10.1080/23808985.2022.2142149

Moran, R. E., & Shaikh, S. J. (2022). Robots in the news and newsrooms: Unpacking meta-journalistic discourse on the use of artificial intelligence in journalism. *Digital Journalism*, *10*(10), 1756–1774. https://doi.org/10.1080/21670811.2022.2085129

Nah, S., McNealy, J., Kim, J. H., & Joo, J. (2020). Communicating artificial intelligence (AI): Theory, research, and practice. *Communication Studies*, *71*(3), 369–372. https://doi.org/10.1080/10510974.2020.1788909

Neijens, P., & Valkenburg, P. M. (2024). How technological and societal developments shaped the agenda of ASCoR. In T. Araujo & P. Neijens (Eds.), *Communication research into the digital society: Fundamental insights from the Amsterdam School of Communication Research* (pp. 33–50). Amsterdam University Press. https://doi.org/10.5117/9789048560592_ch02

Peter, J., Araujo, T., Ischen, C., Shaikh, S. J., van der Goot, M. J., & van Straten, C. L. (2024). Human–machine communication. In T. Araujo & P. Neijens (Eds.), *Communication research into the digital society: Fundamental insights from the Amsterdam School of Communication Research* (pp. 205–220). Amsterdam University Press. https://doi.org/10.5117/9789048560592_ch12

Resendez, V., Araujo, T., Helberger N., & de Vreese, C. (2023). Hey Google! What is in the news? The influence of conversational agents on issue salience. *Digital Journalism*, 1–23. https://doi.org/10.1080/21670811.2023.2234953

Simon, M., Welbers, K., Kroon, A. C., & Trilling, D. (2022). Linked in the dark: A network approach to understanding information flows within the Dutch Telegramsphere. *Information, Communication & Society*. Advance online publication. https://doi.org/10.1080/1369118X.2022.2133549

Sundar, S. S. (2020). Rise of machine agency: A framework for studying the psychology of Human–AI Interaction (HAII). *Journal of Computer-Mediated Communication*, *25*(1), 74–88. https://doi.org/10.1093/jcmc/zmz026

Tanis, M., & Postmes, T. (2003). Social cues and impression formation in CMC. *Journal of Communication*, *53*(4), 676–693. https://doi.org/10.1111/j.1460-2466.2003.tb02917.x

Trilling, D., Araujo, T., Kroon, A., Möller, A. M., Strycharz, J., & Vermeer, S. (2024). Computational communication science in a digital society. In T. Araujo & P. Neijens (Eds.), *Communication research into the digital society: Fundamental insights from the Amsterdam School of Communication Research* (pp. 247–263). Amsterdam University Press. https://doi.org/10.5117/9789048560592_ch15

van den Besselaar, P., & Leydesdorff, L. (1996). Mapping change in scientific specialties: A scientometric reconstruction of the development of artificial intelligence. *Journal of the American Society for Information Science*, *47*(6), 415–436. https://doi.org/10.1002/(SICI)1097-4571(199606)47:63.0.CO;2-Y

Vrijenhoek, S., Kaya, M., Metoui, N., Möller, J., Odijk, D., & Helberger, N. (2021). Recommenders with a mission: assessing diversity in news recommendations. *Proceedings of the 2021 Conference on Human Information Interaction and Retrieval*, 173–183. https://doi.org/10.1145/3406522.3446019

Wang, C., Boerman, S., Moeller, J., & de Vreese, C. H. (2023). *The AI divide: Who is the most vulnerable?* [Manuscript submitted for publication]. Amsterdam School of Communication Research, University of Amsterdam.

About the author

Claes H. de Vreese is Distinguished University Professor of AI and Society with a special emphasis on media and democracy. He co-directs the Gravitation Programme AlgoSoc and the AI, Media and Democracy Lab. He holds the Chair in Political Communication and was Scientific Director of ASCoR from 2005 to 2013. E-mail: C.H.deVreese@uva.nl

14. Person-Specific Media Effects

Patti M. Valkenburg, Ine Beyens, Nadia Bij de Vaate, Loes Janssen, and Amber van der Wal

Abstract

We recently introduced a new approach to study (social) media effects. Our approach challenges the findings of nomothetic media effects studies, which assume that the average effect sizes that they report generalise to all individuals in a (sub-)population. However, using idiographic methods of analysis ($N = 1$ time series analyses), we found striking differences in the person-specific effects of social media use on well-being, ranging from strongly negative ($\beta = -.30$) to strongly positive ($\beta = .35$). Moreover, for only a small minority of respondents, their effect sizes matched with the average effect size of social media use on well-being. Our results show that individuals react and develop in unique ways, and this uniqueness is not captured by approaches that rely on averages.

Keywords: idiographic methods, differential susceptibility, within person changes, experience sampling method (ESM), intensive longitudinal studies

Introduction: From universal to person-specific media effects

Interviewer: What is the influence of social media on you?

Teen 1 (14): Social media helped me discover who I am, what I like and dislike, that way I learn about myself.

Teen 2 (16): Social media changed me in a positive way because now I am more open-minded and down to earth.

Valkenburg, P. M., Beyens, I., Bij de Vaate, N., Janssen, L., & van der Wal, A. (2024). Person-specific media effects. In T. Araujo & P. Neijens (Eds.), *Communication research into the digital society: Fundamental insights from the Amsterdam School of Communication Research* (pp. 233-245). Amsterdam University Press. https://doi.org/10.5117/9789048560592_ch14

Teen 3 (14): On social media you should not compare yourself to others. But that is easier said than done. It has been a difficult time for me. I have been seeing a psychologist for a while now. (van der Wal et al., 2023)

These quotes from Dutch adolescents reflect what media theorists have emphasised for decades: individuals differ strongly in their susceptibility to the effects of (social) media. As early as 1933, the landmark Payne Fund Studies concluded that

> "the movies exert an influence, there can be no doubt. But it is our opinion that this influence is specific for a given child and a given movie. The same picture may influence different children in distinctly opposite directions. Thus in a general survey such as we have made, the net effect appears small" (Charters, 1933, p. 16).

Likewise, in 1948, Berelson, reflecting on what was then known about media effects, concluded that "some kinds of communication on some kinds of issues, brought to the attention of some kinds of people under some kinds of conditions have some kinds of effect" (Berelson, 1948, p. 172).

These sophisticated conclusions in the earliest days of communication research about the boundary conditions of media effects were in part based on case-by-case observations and qualitative interviews. This held for the conclusions of the Payne Fund Studies, but also, for example, for those concerning the effects of radio "daytime serials" on women's perceptions, emotions, and behaviours (Herzog, 1941) and for the analysis of audience reactions to the infamous *War of the Worlds* radio broadcast in 1938 (Cantril, 1952). Studies relying on case-by-case observations or qualitative interviews to uncover differences and communalities in people's responses to their environment are said to employ an idiographic approach. Idiographic studies stem from the idea that each person has unique characteristics that guide their selection of and responses to environmental stimuli, including media (Lerner & Lerner, 2019).

Despite the nuanced conclusions about the effects of media reported in the first half of the 20th century, communication research in the second half of this century has predominantly established universal, across-the-board effects of media that are assumed to hold for all individuals. Such "one effect size fits all" research, which focuses on averages obtained from large samples with the aim to generalise to target populations, is said to employ a nomothetic approach. This historical turn from idiographic conclusions to nomothetic approaches may be due to an increasing use of nomothetic statistics (means, standard deviations, correlations) in the social sciences (Robinson, 2011). Such statistics and their focus on averages have not only dominated empirical research but also meta-analyses, which typically also report only one pooled (and typically small) effect size to summarise certain media effects literatures (Valkenburg et al., 2016).

Media effects theories that appeared in the second half of the 20th century proposed that media effects are conditional, meaning that they differ for individual users. But it was not until the new millennium that attention to differential susceptibility to media effects became the norm in media effects research. As of that time, more comprehensive theories were developed that focused on the cognitive processes that underlie media effects (Lang, 2000), as well as on the dynamic and conditional nature of such effects (Slater, 2007; Valkenburg & Peter, 2013a). From then on, moderator analyses (e.g., multi-group analysis and Hayes' moderated mediation model) rapidly became the standard in the communication discipline. The idea behind this "group-differential approach" is that some characteristics of humans are not shared by all but only by some people (Lerner & Lerner, 2019, p. 27). A group-differential approach allows researchers to investigate which subgroups (e.g., in terms of age or gender) are more (or less) susceptible to the effects of media and technologies.

Despite this undeniable progress in the field, the effect sizes yielded by group-differential approaches are typically also small to moderate at best. Although small effect sizes are common in many disciplines, they are inconsistent with the fine-grained results found in early and recent idiographic studies, which show strong individual differences in responses to (social) media (e.g., van der Wal et al., 2023; Weinstein, 2018). Moreover, small average media effects also defy common sense, because everyday experience offers ample evidence of strong effects for some individuals. A viable explanation for this discrepancy in results between the two approaches may be that group-differential approaches rely on group-level moderators (e.g., gender, age, personality type). Such moderators may invariably gloss over more subtle individual differences in susceptibilities to media effects. As Lerner and Lerner (2019, p. 27) observe, "in addition to their nomothetic and group-differential characteristics, every person has idiographic characteristics that define him or her as unique."

In the past decade, a growing group of methodologists have proposed to unite the subtlety and specificity of idiographic studies with the primary goal of nomothetic studies: drawing conclusions about an underlying population based on samples (Molenaar & Campbell, 2009). Inspired by this methodological work and the new technological opportunities (e.g., phone-based experience sampling tools), our team within ASCoR saw possibilities to introduce such a unified approach to study the idiographic (or $N = 1$) effects of social media use on adolescent well-being. Our new approach, which is currently adopted by several other research teams across the world, combines the strengths of a nomothetic approach (representative samples, generalisation to target populations) with those of an idiographic approach (focus on unique processes within single persons).

How to study $N = 1$: Media effects and what we found

> Interviewer: Hi ChatGPT, can you explain what $N = 1$ media effects research entails?
>
> ChatGPT: Certainly! $N = 1$ or person-specific research is an approach to studying media effects that focuses on understanding how media content affects individuals on a case-by-case basis. Rather than conducting studies with large samples and looking for overall patterns or effects, $N = 1$ researchers collect data from a single individual over an extended period of time, often using methods such as experience sampling.

This information from ChatGPT is remarkably correct. Although not referenced, it must be partly inspired by our work, because until 2021, we were the only research group focusing on person-specific media effects. ChatGPT is right: A person-specific approach relies on data from single persons over an extended period and investigates these persons on a case-by-case basis. In doing so, a person-specific approach acknowledges recent dynamic media effects theories, which consider a media effect as an intra-individual change in cognitions, emotions, or behaviour due to media use, which differs from person to person (Slater, 2007; Valkenburg & Peter, 2013a).

At first sight, the new emphasis on the conceptualisation and investigation of media effects as within-person changes may seem incremental rather than innovative. But one should realise that, until recently, nearly all media effects research has investigated between-person associations of media use with certain outcomes. Between-person and within-person methods of analyses differ fundamentally from each other. Between-person methods try to establish, for example, whether individuals who use more social media are worse off compared to others who use them less. Within-person methods try to uncover whether and how the well-being of persons change when they use more social media than they usually do.

In our and other disciplines, it has long been assumed that between-person results are generalizable to within-person results. But as has been convincingly demonstrated mathematically by Molenaar and Campbell (2009), this assumption is untenable in the social sciences. Our recent experience sampling method (ESM) studies with 126 within-person assessments per adolescent allowed us to investigate to what extent the between-person associations of social media use with well-being differ from the average within-person effects (i.e., the within-person effect across all participants). In addition, our studies allowed us to investigate to what extent the person-specific effects (i.e., the within-person effects for each single person) are consistent with the average within-person effect. In other words, to what extent can we translate the average within-person effect of social media use on well-being to each person in the sample?

We used dynamic structural equation modelling (DSEM) (McNeish & Hamaker, 2020) to examine and compare the between-person, within-person, and person-specific media effects. DSEM is a Bayesian modelling technique that combines the strengths of multilevel analysis and structural equation modelling with $N = 1$ time series analysis. $N = 1$ time series analysis enables researchers to establish the longitudinal effects of media use within a single person. The remaining components of DSEM enables researchers to model the longitudinal effects of multiple persons, while at the same time allowing for group-differential and person-specific differences in media effects.

We indeed found considerable differences in the between-person and the average within-person results. For example, Beyens et al. (2021) found a significant negative between-person association of social media scrolling with well-being ($\beta = -.12$), but no average within-person effect ($\beta = .00$). Sometimes the between- and within-person effects were even opposite to each other, a phenomenon that has been called a Simpson's paradox (Kievit et al., 2013). In our studies, we indeed found support for this paradox. For example, Valkenburg, Beyens, et al. (2022) found a positive between-person association of scrolling with inspiration ($\beta = .08$), but a negative within-person effect ($\beta = -.04$). Likewise, Pouwels et al. (2021) found a positive between-person effect of Instagram use with friends on friendship closeness ($\beta = .17$) but a negative within-person effect ($\beta = -.07$).

We also found significant discrepancies between the average within-person effects and the person-specific effects. The histogram in Figure 14.1 shows the distribution of the person-specific effect sizes of the effect of

Figure 14.1. Distribution of the Person-Specific Effect Sizes of the Effect of Social-Media Scrolling on Well-being

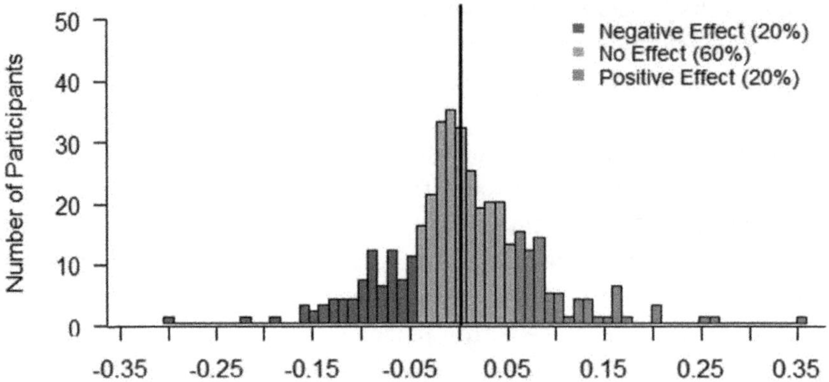

Adapted from Beyens et al., 2021. Results are based on an ESM study among 387 adolescent with 126 within-person assessments across three weeks.

scrolling on well-being found in Beyens et al. (2021). The X-axis of the histogram shows the different person-specific effects sizes of scrolling on well-being, which ranged from moderately negative (β = -.30) to moderately positive (β = +.35). The Y-axis shows the number of participants experiencing specific effect sizes listed on the X-axis. The vertical black line represents the average within-person effect of scrolling on well-being (i.e., β = .00). We found that only 10% of the person-specific effect sizes fell within the credible intervals of the average within-person effect size (β = .00) found in Beyens et al. (2021).

In sum, as demonstrated by Molenaar and Campbell (2009), the average within-person effects indeed systematically differed from the between-person effects of social media use on well-being or related outcomes. In addition, the person-specific effects of only a small minority of respondents were consistent with the average within-person effects. One could argue that variance around an average effect is common and that such variance is, in fact, the basis of stochastic statistics. Indeed, to date most media effect studies, both experimental and correlational, have considered variance around media effects as noise (Valkenburg & Peter, 2013b).

Our unified method allowed us to demonstrate whether the variance around our average within-person effects should be considered as noise or whether individuals truly differ in their responsiveness to media. To do so, we used several accepted methods. For example, in some of our studies we preregistered the smallest effect size of interest (SESOI) (Anvari & Lakens, 2021). In others, we calculated the Bayesian credible intervals for each of the person-specific effects. But while employing these strict methods, we still found that some participants experienced meaningful negative, others meaningful positive, and yet others no effects of social media use on well-being as shown, for example, in Figure 14.1.

We have tried to explain the vast differences in person-specific effects of social media use on well-being and self-esteem. To that end, we have investigated the moderating influences of trait self-esteem, trait self-esteem instability, as well as trait envy, inspiration, and enjoyment in these effects. We have also studied the influence of three time-varying moderators: social media-induced envy, inspiration, and enjoyment (Valkenburg, Beyens, et al., 2022; Valkenburg et al., 2021). The results of these moderating analyses were mixed. We only found that adolescents with an unstable self-esteem and social media-induced envy were more susceptible to negative effects of social media than their peers. But none of these moderators yielded a strong effect, which strengthens us in our notion that each individual has numerous idiosyncratic characteristics that define him/her/them as unique, and that such uniqueness cannot be fully captured by a group-differential approach. To really understand the unique susceptibility factors of individuals, we may need a mixed-mode approach, in which $N = 1$ time series are combined with in-depth interviews with participants.

Contributions to theory and practice

An important contribution of our person-specific media effects approach is that it introduces a new way to investigate the validity of hypotheses about media effects in populations or sub-populations. More specifically, our approach allows us to reveal for how many participants a media effects hypothesis is confirmed and for how many not. For Karl Popper (1959), the observation of a single case that did not conform to the hypothesis would be enough to falsify a hypothesis. In our work, we argue that a media effects hypothesis is valid only if it applies to the vast majority of participants (i.e., >75%).

Recently, we investigated a recurrent hypothesis in the literature, the "passive social media use hypothesis" (Verduyn et al., 2017). This hypothesis states that passive social media use (i.e., browsing/scrolling) results in lower well-being, because it leads to upward social comparison and envy, which in turn negatively affects one's well-being. Despite its appeal, a recent systematic review on the passive social media use hypothesis yielded large inconsistency in the included studies, meaning that some studies found support for the hypothesis, while others did not (Valkenburg, van Driel, et al., 2022).

We believe that these inconsistencies can be explained by two factors: person-specific susceptibilities and sampling procedures. Our studies showed that individuals differ strongly in their responses to social media. We found that the passive social media use hypothesis was confirmed for only 20% of participants, while it was not supported for 80% of participants (the red bars on the left of Figure 14.1). We also found that 20% of participants even experienced an effect opposite to the hypothesis (the green bars on the right in Figure 14.1) (Beyens et al., 2021). Thus, to speak in Popperian terms, our approach enabled us to falsify the passive social media use hypothesis.

Our approach not only has theoretical relevance, but also great societal implications. For example, take a moment to think about prevention and intervention programmes. Our knowledge could be of vital importance for the development of such programmes. After all, if practitioners would base such programmes on average results, they may conclude that such programmes are not necessary, because most studies, including our own, report only very small average effects of social media. But a considerable minority of adolescents did experience meaningful negative effects of social media use. Based on our results, it may be safe to assume that 10% of young people experience negative effects of social media use. And knowing that the Netherlands has 1 million and the US 75 million young people, these negative effects may generalise to 100,000 Dutch and 7.5 million US minors. These young people may run the risk of experiencing mental health problems due to their social media use. We cannot deny that we need to take such percentages seriously.

Our research may have a wide range of implications in various fields, ranging from social media-induced mental health to advertising and political communication. Person-specific studies could be used to develop personalised interventions, for example, to identify specific triggers of mental health problems in participants or patients. It can be used to study the effectiveness of entertainment or advertising campaigns to understand what content or messages are most effective for specific individuals or subgroups. Finally, it can be used to investigate the effect of political communication and campaigns on a case-by-case basis. Researchers could track individual responses to different types of political messages and identify which messages are most influential among specific individuals or subgroups.

Next steps

In this chapter, we made a case for a person-specific approach to study the effects of media, that is, a focus on within-person processes that differ from person to person. We do not deny that between-person methods of analysis are important, for example, to compare subgroups or generations in terms of their preferences for political parties, their well-being, or their consumer behaviour. But based on our findings, we do believe that between-person methods of analysis are sub-optimal to study media effects. Cross-sectional correlational studies are inherently unable to establish within-person changes, but many early longitudinal studies have not been able to do so either. It is only as of 2017, when Random-Intercept and multi-level models gradually became the norm in our discipline, that we are able to investigate over-time within-person media effects. Even though cross-sectional "effect studies" still appear in communication journals, the time is ripe to agree that such studies are not suited to investigate media effects. Cross-sectional and other between-person designs overlook a principal proposition of media effects theories, namely that a media effect is a within-person change process that differs across persons.

Although our team was the first to introduce a person-specific approach into the communication discipline, we must acknowledge that as academics we are greatly lagging behind when we compare our knowledge to that of tech companies and social media platforms. Without exception, these companies and platforms conduct $N = 1$ research on the preferences of each single user, which are used to improve their recommendation systems, to keep each of us glued to our personalised screens, and to sell their person-specific preferences to advertisers. And we all give them our consent to do so.

Whereas the experiences of Facebook and Instagram users were mainly driven by social interactions, newer platforms such as TikTok offer their users algorithmically driven experiences at the $N = 1$ level, meaning that each single user can create their own experiences, and thereby possibly their own effects (Bhandari & Bimo, 2022). Until now, knowledge of such idiographic user experiences is limited to these tech companies. An important step for future academic research is to combine our $N = 1$ research methods with these algorithmically based user experiences. There are various methods to capture such experiences. A discussion of these methods is beyond the scope of this chapter, but an up-to-date and comprehensive review of such methods can be found in Ohme et al. (2023).

In this chapter, we have attempted to disconfirm the fallacy of the "one effect size fits all" approach that has characterised the communication research tradition for too long. Our results show that small average media effects can have important implications for some media users. Our results imply that the small average statistics that have been reported for decades in both empirical research and meta-analyses do not do justice to the minorities of individuals who benefit from media use and to other minorities who suffer from media use. We agree with Rose et al. (2013) that individuals think, behave, and develop in distinctive ways, thereby showing variability that is not captured by statistics that solely rely on averages. We communication scholars "have a bright future before us, and it begins where the average ends" (Rose, 2016, p. 191).

References

Anvari, F., & Lakens, D. (2021). Using anchor-based methods to determine the smallest effect size of interest. *Journal of Experimental Social Psychology, 96*, Article 104159. https://doi.org/10.1016/j.jesp.2021.104159

Berelson, B. (1948). Communication and public opinion. In W. Schramm (Ed.), *Communications in modern society* (pp. 167–185). University of Illinois Press.

Beyens, I., Pouwels, J. L., van Driel, I. I., Keijsers, L., & Valkenburg, P. M. (2021). Social media use and adolescents' well-being: Developing a typology of person-specific effect patterns. *Communication Research*. Advance online publication. https://doi.org/10.1177/00936502211038196

Bhandari, A., & Bimo, S. (2022). Why's everyone on TikTok now? The algorithmized self and the future of self-making on social media. *Social Media + Society, 8*(1), 1–11. https://doi.org/10.1177/20563051221086241

Cantril, H. (1952). *The invasion from Mars: A study in the psychology of panic*. Princeton University Press.

Charters, W. W. (1933). *Motion pictures and youth: A summary*. Macmillan. https://ia601608.us.archive.org/4/items/motionpicturesyo00charrich/motionpicturesyo00charrich.pdf

Herzog, H. (1941). On borrowed experience. *Studies in Philosophy and Social Science, 9*(1), 65–95.

Kievit, R., Frankenhuis, W., Waldorp, L., & Borsboom, D. (2013). Simpson's paradox in psychological science: A practical guide. *Frontiers in Psychology, 4*, Article 513. https://doi.org/10.3389/fpsyg.2013.00513

Lang, A. (2000). The limited capacity model of mediated message processing. *Journal of Communication, 50*(1), 46–70. https://doi.org/10.1111/j.1460-2466.2000.tb02833.x

Lerner, R. M., & Lerner, J. V. (2019). An idiographic approach to adolescent research: Theory, method, and application. In L. B. Hendry & M. Kloep (Eds.), *Reframing adolescent research* (pp. 25–38). Routledge. https://doi.org/10.4324/9781315150611-2

McNeish, D., & Hamaker, E. L. (2020). A primer on two-level dynamic structural equation models for intensive longitudinal data in Mplus. *Psychological Methods, 25*(5), 610–635. https://doi.org/10.1037/met0000250

Molenaar, P. C. M., & Campbell, C. G. (2009). The new person-specific paradigm in psychology. *Current Directions in Psychological Science, 18*(2), 112–117. https://doi.org/10.1111/j.1467-8721.2009.01619.x

Ohme, J., Araujo, T., Boeschoten, L., Freelon, D., Ram, N., Reeves, B. B., & Robinson, T. N. (2023). Digital trace data collection for social media effects research: APIs, data donation, and (screen) tracking. *Communication Methods and Measures.* Advance online publication. https://doi.org/10.1080/19312458.2023.2181319

Popper, K. (1959). *The logic of scientific discovery.* Hutchinson & Co.

Pouwels, J. L., Valkenburg, P. M., Beyens, I., van Driel, I. I., & Keijsers, L. (2021). Social media use and friendship closeness in adolescents' daily lives: An experience sampling study. *Developmental Psychology, 57*(2), 309–323. https://doi.org/10.1037/dev0001148

Robinson, O. C. (2011). The idiographic/nomothetic dichotomy: Tracing historical origins of contemporary confusions. *History & Philosophy of Psychology, 13*(2), 32–39. https://doi.org/10.53841/bpshpp.2011.13.2.32

Rose, L. T., Rouhani, P., & Fischer, K. W. (2013). The science of the individual. *Mind, Brain, and Education, 7*(3), 152–158. https://doi.org/10.1111/mbe.12021

Rose, T. (2016). *The end of average: How to succeed in a world that values sameness.* HarperCollins.

Slater, M. D. (2007). Reinforcing spirals: The mutual influence of media selectivity and media effects and their impact on individual behavior and social identity. *Communication Theory, 17*(3), 281–303. https://doi.org/10.1111/j.1468-2885.2007.00296.x

Valkenburg, P. M., Beyens, I., Pouwels, J. L., van Driel, I. I., & Keijsers, L. (2022). Social media browsing and adolescent well-being: Challenging the "passive social media use hypothesis." *Journal of Computer-Mediated Communication, 27*(1), 1–19. https://doi.org/10.1093/jcmc/zmab015

Valkenburg, P. M., & Peter, J. (2013a). The differential susceptibility to media effects model. *Journal of Communication, 63*(2), 221–243. https://doi.org/10.1111/jcom.12024

Valkenburg, P. M., & Peter, J. (2013b). Five challenges for the future of media-effects research. *International Journal of Communication, 7*, 197–215. http://ijoc.org/index.php/ijoc/article/download/1962/849

Valkenburg, P. M., Peter, J., & Walther, J. B. (2016). Media effects: Theory and research. *Annual Review of Psychology, 67*, 315–338. https://doi.org/10.1146/annurev-psych-122414-033608

Valkenburg, P. M., Pouwels, J. L., Beyens, I., van Driel, I. I., & Keijsers, L. (2021). Adolescents' experiences on social media and their self-esteem: A person-specific susceptibility perspective. *Technology, Mind, and Behavior, 2*(2). https://doi.org/10.1037/tmb0000037

Valkenburg, P. M., van Driel, I. I., & Beyens, I. (2022). The associations of active and passive social media use with well-being: A critical scoping review. *New Media & Society, 24*(2), 530–549. https://doi.org/10.31234/osf.io/j6xqz

van der Wal, A., Valkenburg, P. M., & van Driel, I. I. (2023). In their own words: How adolescents differ in their social media use and how it affects them. PsyArxiv. https://doi.org/10.31234/osf.io/mvrpn

Verduyn, P., Ybarra, O., Résibois, M., Jonides, J., & Kross, E. (2017). Do social network sites enhance or undermine subjective well-being? A critical review. *Social Issues and Policy Review, 11*(1), 274–302. https://doi.org/10.1111/sipr.12033

Weinstein, E. (2018). The social media see-saw: Positive and negative influences on adolescents' affective well-being. *New Media & Society, 20*(10), 3597–3623. https://doi.org/10.1177/1461444818755634

About the authors

Patti M. Valkenburg is a Distinguished University Professor at ASCoR. Her research focuses on the social and emotional effects of (social) media on youth and adults. She is particularly interested in theorising, studying, and demonstrating how individuals differ in their susceptibility to the effects of (social) media. E-mail: P.M.Valkenburg@uva.nl

Ine Beyens is an Assistant Professor of Youth & Media Entertainment at ASCoR. Her research focuses on the effects of screen media on the psychological and social well-being of adolescents. In her work, she employs intensive longitudinal data collected through experience sampling methods (ESMs) to capture how adolescents use screen media in their daily life. E-mail: I.Beyens@uva.nl

Nadia Bij de Vaate is a Postdoctoral Researcher at ASCoR and a member of project AWeSome (Adolescents, Well-being, and Social media). Her research interests include how adolescents shape who they are and how they feel resulting from the emergence and continuation of social media usage. E-mail: A.J.D.Bijdevaate@uva.nl

Loes Janssen is a Postdoctoral Researcher at ASCoR. Her research focuses on the effects of social media and parent–adolescent communication on the well-being of adolescents. In her work, she employs intensive longitudinal data to capture the daily life experiences of adolescents. E-mail: L.H.C.Janssen@uva.nl

Amber van der Wal is a Postdoctoral Researcher at ASCoR and a member of project AWeSome (Adolescents, Well-being, and Social media). She investigates the effects of social media use on the well-being of adolescents, with a special focus on the role of humour in this process. E-mail: A.vanderWal@uva.nl

15. Computational Communication Science in a Digital Society

Damian Trilling, Theo Araujo, Anne Kroon, A. Marthe Möller, Joanna Strycharz, and Susan Vermeer

Abstract

Computational methods have added new approaches to the way many communication scientists do their work. We identify four developments that accelerated the adaption of computational methods: the increasing availability of digital data, the surge of large amounts of user-created data, the need to study new artefacts, and the improved accessibility of computational resources. We describe new data acquisition techniques, new research designs, and new analytical approaches that characterise the field. After discussing contributions to the open source community, to the methodological toolbox, as well as to the testing and development of theories, we sketch in broad strokes a research agenda for the coming years.

Keywords: computational communication science, computational social science, digital society, computational methods

Introduction

For a long time, communication scientists have used computers to aid their work. By the 1970s, if not earlier, some communication scientists had built computer models that searched for keywords in texts to conduct automated content analyses of news (Schönbach, 1982). Yet, only recently, such computational approaches became part of the communication research mainstream. For instance, around 15 years ago, Lazer et al. (2009) popularised the term "computational social science." Analogously, nowadays, "computational communication science" can be considered an established term

Trilling, D., Araujo, T., Kroon, A., Möller, A. M., Strycharz, J., & Vermeer, S. (2024). Computational communication science in a digital society. In T. Araujo & P. Neijens (Eds.), *Communication research into the digital society: Fundamental insights from the Amsterdam School of Communication Research* (pp. 247-263). Amsterdam University Press. https://doi.org/10.5117/9789048560592_ch15

(e.g., Hilbert et al., 2019), with its own division (Computational Methods) at the International Communication Association and a dedicated journal (*Computational Communication Research*) (van Atteveldt et al., 2019).

While it may be too early for a clear consensus within the discipline about the exact characteristics and boundaries of computational communication science, for the purpose of this contribution, we follow Hilbert et al., who "define computational communication science as the endeavor to understand human communication by developing and applying digital tools that often involve a high degree of automation in observational, theoretical, and experimental research" (2019, p. 3915). This definition highlights a number of important aspects: First, neither the objective (understanding human communication) nor the general methodological approaches (theoretical, observational, or experimental work) has changed—we are still doing communication science. Second, to do so, we develop and apply new tools, allowing us to work on a different scale than previously possible.

We observe four intertwined developments that lead to the increasing attention to computational approaches. First, the data that communication scientists are interested in are increasingly available in digital formats. News sites have started to replace printed newspaper editions, an analogue telephone call or chat in a pub may now be observable on social media, and also advertising has often gone digital.

Second, whereas media content typically was created by "institutions" such as newspapers, corporate organisations, or governments, nowadays, media users play an important role in the creation of data, impacting their amount, content, and ownership. Traditional techniques may not scale well enough to collect and analyse such data.

Third, new artefacts such as voice assistants, search engines, chatbots, or recommender systems have become relevant for people's media use. Some of the aspects of such artefacts require new methodological approaches to study them. On a higher level, the increased personalisation of the media environment also falls under this development: There is no longer one newspaper, one advertisement, or one television programme that is largely saying the same thing for (in principle) everyone. Instead, individuals receive different information about different topics at different moments in time. Both the collection and the analysis of such personalised data requires—at least to an extent—automation.

Fourth and finally, advances in and accessibility of storage, computing power, but also powerful and more accessible programming frameworks have enabled things that before would have required immense resources in terms of finances and expertise.

Due to these developments, researchers at our department and beyond have turned to computational methods to study communication with increasing frequency. This chapter provides an overview of how this impacts the work of communication scholars and its implications for the field. As the boundaries of what constitutes computational communication science may sometimes be a bit blurry, our list may not be exhaustive. For instance, our examples tend to emphasise content-analytical and digital trace data, while other types of computational work (such as theory-driven computer simulations) have only slowly been taken up in our department so far. Yet, we believe it covers the main developments we observe at this moment in the work done at ASCoR.

Empirical findings

As we have seen, the boundary between traditional and computational communication science is fluid and, sometimes, it is just about automating and/or scaling up one part of the research pipeline. In this section, we discuss three main parts in which computational techniques are employed: data acquisition, research design, and analytical approaches.

New data acquisition techniques

The collection of digitally available data, such as online news, social media content, or other types of digital traces, requires new methods. Three main paths have been used at ASCoR to collect such data: (1) application programming interfaces (APIs); (2) scraping; and (3) tracking and data donation, that is, partnering with users willing to install trackers or to donate their data to academic research.

Application programming interfaces (APIs). Some platforms and services allow scholars to gather data via APIs. These services, originally developed to enable third-party clients to interact with a platform in a standardised format, sometimes also allow searching for and downloading specific data subsets. Over the years, many ASCoR researchers have been using this functionality to collect large amounts of social media data, for example, posts or comments from Facebook, Twitter, or YouTube. While application programming interfaces (APIs) provide researchers with significant benefits, such as structured access to extensive data, it is important to note their limitations. Notably, the availability of API access has been increasingly restricted as platforms impose regulations on data accessibility. For example,

Twitter introduced several restrictions that first required academics to apply for full access to the data and now created a prohibitively expensive paid tier to the API that also applies to academics. This makes APIs provided by platforms an unsustainable data collection practice that cannot be trusted to be available and necessitates researchers to turn to different ways of collecting data.

Scraping. In cases when no or very limited access through API is provided by a platform, online content has been commonly scraped. This data collection method involves crawling a website with an automated script that explores the website's structure and collects data using HTTP requests and responses. At ASCoR, scraping has, for instance, routinely been used for conducting content analyses of Dutch online news by scraping news websites at a regular interval. To do so, scholars have collaborated and shared the infrastructure for such content analyses (e.g., Trilling et al., 2018). Scraping has also been used to collect social media content not available via APIs, for example, using open source scrapers to collect visual posts from Instagram or TikTok. While independent from proprietary access given by platforms, it presents a challenge in balancing data collection efforts with currently often unclear regulations that govern scraping different websites.

Tracking and data donation. The usage of APIs or scraping to collect digital trace data (often from social media) enables the creation of large-scale datasets of communication data, yet it often is limited to publicly available data (e.g., Instagram posts that are public), subject to restrictions by platforms, and does not allow for the linkage to individual-level information about the causes (e.g., motivations) or consequences (e.g., perceptions, attitudes) of communication. To address these challenges, ASCoR researchers have extended or developed approaches to tracking and data donation, in which researchers partner with participants willing to share their data. Studies have been conducted with data gathered by apps or trackers that individuals install on their devices, enabling, for example, examining the way news consumers navigate online (Möller et al., 2020; Vermeer et al., 2020) or assessing the differences in self-reported and tracked internet or social media use (Araujo et al., 2017; Verbeij et al., 2021). Another approach is for individuals to request their data from online platforms, and donate it to researchers. ASCoR researchers have not only developed an open source framework for data donation (Araujo et al., 2022), but also several research groups within the institute have created guidelines and methodological studies on the advantages, limitations, and the potential of the method (van Driel et al., 2022). Others have relied on donations of local browsing histories from participants' web browsers (Wojcieszak et al., 2022), analysed participants' WhatsApp data (Vermeer et al., 2021), combined data donations with mobile experience sampling, in which participants get multiple short questions on their smartphone within a short time frame (Otto & Kruikemeier, 2023), or asked participants to donate recordings of the iOS screen time function (Ohme et al., 2020).

New research designs

The increasing digitalisation and datafication of the communication landscape not only enables new forms of collecting communication data for research, but also brings forward a new set of artefacts that, in and by themselves, become the object of communication research. To give an example: Communication scientists can use machine learning to analyse media content—but they can also study how media companies employ machine learning and analyse its consequential impact. Accordingly, as search engines and recommender systems become increasingly central in the media landscape, ASCoR researchers devised new research designs to experimentally study the causes, contents, and consequences of individual interactions with these artefacts.

As content that users are exposed to online is increasingly personalised, ASCoR researchers have turned to designing their own prototypes of so-called recommender systems used for such personalisation to be able to closely observe how individuals interact with them and link these interactions to motivations and consequences. Integrating such prototypes in experimental research allows the increase of ecological validity going beyond scenario studies used in the past. As personalisation is currently present in all domains of communication, such research has been conducted on the impact of recommendation on diversity of news consumption (Loecherbach et al., 2021), possibilities for increasing effectiveness of health information (Nguyen et al., 2017), or achieving a healthy lifestyle through a personalised mobile coaching app (Stuber et al., 2020).

In addition, ASCoR researchers have deployed new research designs to study the emergence of conversational agents (e.g., chatbots and virtual assistants) and social robots in the communication environment. For an overview, see Peter et al. (2024) in this book.

New analytical approaches

New types of data often require new analytical approaches. We saw that already quite early, ASCoR researchers started analysing, for instance, word frequencies. For example, pioneered by Loet Leydesdorff, colleagues examined co-occurrence matrices of words to study topics and frames (e.g., Hellsten et al., 2010; Leydesdorff & Welbers, 2011). Such bottom-up analyses were complemented with top-down dictionary approaches, in which colleagues compiled lists of keywords to automatically search for pre-defined topics, frames, and sentiments.

In contrast, contemporary analyses at ASCoR typically make use of machine learning approaches to, for example, identify frames in news articles, categorise their topic, or to estimate the sentiment of social media data. This means that the top-down dictionary approach (in which researchers come up with a list of keywords that are then automatically counted), in many instances has been replaced by supervised machine learning: We hand-code a subset of the data and estimate a model that can predict the coding based on word frequencies. In other words, and at the danger of oversimplifying matters, rather than pretending that we can come up with a perfect algorithm ourselves, we just let the computer "learn" an algorithm (and associated weights) from a sample of the data. Regarding bottom-up approaches, besides classic techniques like cluster analysis or principal component analysis, unsupervised approaches such as so-called topic models have become popular to identify topics and frames that are not known a priori. In short: supervised machine learning uses hand-coded (labelled) data to learn how to automatically code a larger (unlabelled) dataset, while unsupervised machine learning does not use hand-coded data and recognises patterns in the data instead.

While such classic machine learning approaches (both supervised and unsupervised) are still popular, recent work increasingly focuses on techniques that go beyond this classic approach (in which word order is not taken into account, and in which we cannot account for similarities between words, such as synonyms or antonyms). Researchers now regularly work with word embeddings (vector representations of words, in which more similar words are closer to each other in a vector space), which are either studied in themselves (for instance, to detect biases), or made part of a supervised or unsupervised machine-learning workflow. In particular, researchers work with so-called neural networks. In such a network, features (such as word frequencies) do not directly predict the outcome (such as the topic of a text) anymore but predict some intermediate layers first. Once the number of layers grows, such an approach is also called deep learning. Crucially, deep learning makes it possible to move beyond text: We can as well, for example, use pixels of an image as input features, and the different layers than "learn" how, for instance, the furriness of a cat may predict that it is not a dog.

Very recently, though, so-called transformer-based approaches have revolutionised the field. Even lay people have heard about large language models like ChatGPT, which can generate text based on free-form text prompts, or DALL-E, which can generate images based on text prompts. ASCoR researchers not only study such models (for example, to understand the biases they contain), but also employ them in their research. For example, we learned that neural topic models, which do not "start at zero" but can build on existing language models, provide much more coherent topics than the models that have been used so far. And fine-tuning an existing model by providing it with additional information (e.g., annotations of relevant texts) such that it learns a supervised classification task that we are interested in, often (but not always) outperforms classic approaches and/or requires much fewer training examples (so-called few-shot learning). On the other hand, such approaches also require specific computing resources (graphics processing units, or GPUs), accessibility to which provides a barrier to ASCoR researchers at the moment.

Contributions to the field

So far, we have outlined the developments that have led to innovations within the research done at ASCoR. We now look at the next step and focus not so much on what has been *used* in our group, but about what our group has *provided*. These contributions are not only relevant because they advance the development of practical tools that scholars can use, but also because they can be used to advance our theoretical understanding of communication processes and their consequences, or because they can be used to test theories that were not testable before. Next to the contributions to the computational communication science field, which we outline in the next paragraphs, through intense collaboration with the University of Amsterdam's Institute for Information Law (IViR), lines of research outlined in this chapter also translate into specific policy recommendations.

Contributions to the open source community

Many of the new approaches described above have led to the development of new open source tools that can and have been used for follow-up studies. For example, Araujo (2020) developed a toolkit that offers researchers the possibility of integrating a conversational agent into their experiments; Loecherbach and Trilling (2020) developed a tool to create online experiments with news recommender systems; Votta (e.g., 2021) created multiple R packages to make computational tools more easily accessible; and a large group of colleagues developed OSD2F, an open source data donation framework that allows anybody to set up privacy-respecting data donation studies (Araujo et al., 2022). In sum, these contributions do not only allow gaining access to more and more varied data, but have also increased the possibilities for data creation. Specifically, integrating computational methods into, for example, established research designs (such as social-scientific experiments) has opened a range of new opportunities to conduct studies, often in a more ecologically valid way than previously possible.

Validation and consolidation of the toolbox

New methodological approaches beg the question: How valid and reliable are they—and which ones stand the test of time and should be added to our toolbox? For *new research designs*, multiple efforts have been undertaken, for instance, to understand survey measures of media use vis-à-vis digital trace data. For example, this is done by Valkenburg and her team, who compared digital trace data with survey data gathered among social media users (Verbeij et al., 2021), or by another group focusing on internet use (Araujo et al., 2017). For *new analytical approaches*, others have focused on assessing the quality of such automated content analysis techniques—for instance, the work of Boukes et al. (2020) showing the poor performance of off-the-shelf sentiment analysis dictionaries. The estimates provided by such dictionaries turned out to correlate very weakly with each other and with human coders. And, indeed, also thanks to studies such as theirs, the technique is now considered outdated by many. Finally, much time has been invested (not only in research, but also in teaching) to consolidate and canonise the techniques—for instance, for automated content analysis (Boumans & Trilling, 2016)—that culminated in the publication of a textbook (van Atteveldt et al., 2022).

New avenues for developing and testing theories

Using computational techniques to gather, generate, and analyse data allows scholars to advance their theoretical understanding of communication processes and their effects. The rise of the digital society has led some established paradigms within the field to be questioned and using computational methods in their research equips scholars to understand these phenomena better. Some examples may illustrate such relationships between methodological and theoretical innovations. One example is the acceleration of research on artificial entities in our communication environment—when we see the medium as the communication partner (Peter & Kühne, 2018; Zhao, 2006)—with new methods allowing for chatbot experiments (e.g., Araujo, 2020) helping further develop and test such theoretical approaches.

Scraping in combination with advanced large-scale text comparison approaches and network clustering approaches allow the modification of agenda-setting theory such that it can be applied to a fine-grained news event level, rather than to the comparatively coarse issue level that agenda-setting theory was confined to due to feasibility constraints (Trilling & van Hoof, 2020). Similarly, combinations of scraping, social media APIs, network clustering, and different automated content analysis techniques can help us improve theories about information dissemination and opinion leaders, both regarding brand content (e.g., Himelboim et al., 2023) as well as political content (e.g., Simon et al., 2022). Tracking data (e.g., Merten et al., 2022) are also used for that purpose.

Theories about the political role of interpersonal talk—also hard to test in an offline setting—can now be refined by observing political talk in app data (e.g., Vermeer et al., 2021). The use of large language models (LLMs) now makes it possible to reconcile journalistic genre theories with an analysis of extremely varied and multi-modal news-related online content (Lin et al., 2023). Not only are technological developments changing media content consumption, but they also impact individuals' knowledge and experience with various technologies, leading to new theoretical areas to explore. Focusing on advertisements, in particular, Strycharz and Segijn (2022) describe how the constant collection and usage of consumer data for advertising purposes influence consumer perceptions and behaviour. Such examples may offer a glimpse of the theoretical challenges that the digital society brings with it—but also of the potential that computational methods can have when it comes to developing and testing such theoretical innovations.

Next steps

So far, we outlined multiple promising developments in the field of computational communication science, and it is to be expected that work on all of them will continue. Nevertheless, at the risk of omitting important items, this concluding section provides an attempt to highlight some very recent developments that will most likely shape our research profoundly in the next few years.

Expand research across content domains, languages, and modalities

In recent years, the field of natural language processing (NLP) has been revolutionised by the development of LLMs based on deep learning architectures. LLMs have the potential to significantly augment the efficiency and capabilities of computational analysis applied to media texts in various content domains (including political texts and social media posts) and languages (e.g., Lin et al., 2023). Currently, they are reshaping the benchmarks and standards for automated content analysis. In inductive research, LLMs, powered by tools such as BERTopic, can aid in discovering hidden patterns in text and clustering content around topics, actors, or sentiment (e.g., Simon et al., 2022). For deductive research, pre-trained LLMs can be fine-tuned on small-scale annotated custom datasets to improve the accuracy of concept classification, such as frames, topics, sentiment, or incivility, which are central to communication science. The multilingual and multi-modal analysis capabilities of LLMs are particularly promising, and ASCoR researchers are currently exploring these avenues to enable cross-lingual analyses in different content domains, even including visual and textual content in the same models.

Advance understanding of the impact of algorithmic systems

The advancement of computational methods and the development of LLMs will likely add to the importance of content-based recommendations (over other techniques, such as collaborative filtering)—making such models especially interesting from a research perspective. Consequently, further research on this topic at ASCoR is crucial. Despite the significant performance gains that LLMs bring to the field of computational communication science, there are also concerns about their potential negative influence on research and society at large. A first, salient concern is that LLMs may resonate, reflect, and reinforce existing (historical) biases and inequalities in society. These models are typically trained on large amounts of textual data derived from a broad spectrum of online sources, generated by humans. These data are by no means neutral but reflect human preferences and biases. These biases in the training data may subsequently become embedded in the inner workings of LLMs and produce predictions that are biased in implicit ways. While such biases may be difficult to detect, they can affect society at scale. In addition, there are concerns about the lack of diversity among developers, which may lead to algorithmic design choices that do not adequately address the concerns and input of underrepresented groups. Consequently, biases may become "engineered" in LLMs due to algorithmic design choices and biased training data. This may have consequences for the classification performance of fine-tuned models but may also prompt biases responses in machine-generated communication, such as that created by chatbots. Given the societal importance of these issues, ASCoR researchers have an interest in studying them. This work will build upon previous research on bias in language models conducted by the group's researchers (e.g., Kroon et al., 2021), and focus more in depth on bias in LLMs and consequences for downstream classification. Another significant concern about LLMs relates to their contribution to the so-called authenticity crisis of communication, as it becomes progressively more difficult to distinguish real or authentic communication from fake or unauthentic. There are significant concerns about the ability of LLMs to be (mis)used to generate fake news or ads, spread disinformation, and impersonate real or even fake individuals. This concern has already inspired research by members of ASCoR, for example, on the implications of so-called deepfakes (Dobber et al., 2021). Future research by members of our group will be inspired by these themes.

Computational methods as facilitators of human–machine communication

With the development of methods such as LLMs, computers' ability to "understand" and generate natural language makes them ideal for use in human–machine communication, particularly for conversational agents. As machine-generated communication becomes increasingly difficult to distinguish from human communication, LLMs can inspire and support research at ASCoR into the consequences of conversational agent feature designs and communication styles. These advances can create interesting theoretical questions that draw on state-of-the-art computational techniques. In addition, LLMs are valuable in researching recommender systems, as they can cluster articles that discuss similar topics and/or express similar sentiments towards specific political attitudes, topics, and actors. This clustering enhances recommender systems' ability to personalise content for individuals' prior political attitudes and beliefs, with significant consequences for democratic functioning (e.g., Dobber et al., 2021).

Exposure to and dissemination of information in the digital society

A central theme in communication science is how information spreads in societies. But while in earlier days, for example, agenda-setting studies could suffice with studying relatively coarse issues in a handful of outlets—but in an environment in which not only traditional media, but also social media, alternative media, and fringe platforms outside of the mainstream (such as Telegram) play a role, we need new approaches—both theoretically and methodologically. A lot of work ahead in this area, in particular, devising methodological approaches that combine multiple techniques: APIs, scraping and data donations for data collection, machine learning based on LLMs in combination with network analysis for the identification of topics, themes, and events, and—additionally—simulation-based approaches to model complex non-linear relationships.

These new data acquisition techniques, new research designs, and new analytical approaches open several promising avenues for communication research in the digital society. Advancing our understanding of (the performance of) computational techniques and using this understanding to further develop these techniques are essential steps in providing scholars with more possibilities to analyse the fast quantity of various communication data generated by individuals, groups, and organisations in the digital society. Given the complexity and fast development of these techniques, ASCoR established the Digital Communication Methods Lab (digicomlab), in 2018.[1] It is a place where ASCoR scholars interested in and working with digital methods come together to address these challenges, and to collaborate with the communication science community more broadly.

The road ahead is still long and, as we have shown, not without risk. But ultimately, to understand communication in the digital society, we will need to push ahead and embrace innovative tools that will allow us to study what we need to study. We are working on it.

References

Araujo, T. (2020). Conversational Agent Research Toolkit: An alternative for creating and managing chatbots for experimental research. *Computational Communication Research, 2*(1), 35–51. https://doi.org/10.5117/CCR2020.1.002.ARAU

Araujo, T., Ausloos, J., van Atteveldt, W., Loecherbach, F., Moeller, J., Ohme, J., Trilling, D., van de Velde, B., de Vreese, C., & Welbers, K. (2022). OSD2F: An open-source data donation framework. *Computational Communication Research, 4*(2), 372–387. https://doi.org/10.5117/CCR2022.2.001.ARAU

1 https://www.digicomlab.eu.

Araujo, T., Wonneberger, A., Neijens, P., & de Vreese, C. (2017). How much time do you spend online? Understanding and improving the accuracy of self-reported measures of internet use. *Communication Methods and Measures*, *11*(3), 173–190. https://doi.org/10.1080/19312458.2017.1317337

Boukes, M., van de Velde, B., Araujo, T., & Vliegenthart, R. (2020). What's the tone? Easy doesn't do it: Analyzing performance and agreement between off-the-shelf sentiment analysis tools. *Communication Methods and Measures*, *14*(2), 83–104. https://doi.org/10.1080/19312458.2019.1671966

Boumans, J. W., & Trilling, D. (2016). Taking stock of the toolkit: An overview of relevant automated content analysis approaches and techniques for digital journalism scholars. *Digital Journalism*, *4*(1), 8–23. https://doi.org/10.1080/21670811.2015.1096598

Dobber, T., Metoui, N., Trilling, D., Helberger, N., & de Vreese, C. (2021). Do (micro-targeted) deepfakes have real effects on political attitudes? *The International Journal of Press/Politics*, *26*(1), 69–91. https://doi.org/10.1177/1940161220944364

Hellsten, I., Dawson, J., & Leydesdorff, L. (2010). Implicit media frames: Automated analysis of public debate on artificial sweeteners. *Public Understanding of Science*, *19*(5), 590–608. https://doi.org/10.1177/0963662509343136

Hilbert, M., Barnett, G., Blumenstock, J., Contractor, N., Diesner, J., Frey, S., González-Bailón, S., Lamberso, P., Pan, J., Peng, T.-Q., Shen, C., Smaldino, P. E., van Atteveldt, W., Waldherr, A., Zhang, J., & Zhu, J. J. H. (2019). Computational communication science: A methodological catalyzer for a maturing discipline. *International Journal of Communication*, *13*, 3912–3934. https://ijoc.org/index.php/ijoc/article/view/10675

Himelboim, I., Maslowska, E., & Araujo, T. (2023). Integrating network clustering analysis and computational methods to understand communication with and about brands: Opportunities and challenges. *Journal of Advertising*. Advance online publication. https://doi.org/10.1080/00913367.2023.2166629

Kroon, A. C., Trilling, D., & Raats, T. (2021). Guilty by association: Using word embeddings to measure ethnic stereotypes in news coverage. *Journalism & Mass Communication Quarterly*, *98*(2), 451–477. https://doi.org/10.1177/1077699020932304

Lazer, D., Pentland, A., Adamic, L., Aral, S., Barabási, A.-L., Brewer, D., Christakis, N., Contractor, N., Fowler, J., Gutmann, M., Jebara, T., King, G., Macy, M., Roy, D., & van Alstyne, M. (2009). Computational social science. *Science*, *323*(5915), 721–723. https://doi.org/10.1126/science.1167742

Leydesdorff, L., & Welbers, K. (2011). The semantic mapping of words and co-words in contexts. *Journal of Infometrics*, *5*(3), 469–475. https://doi.org/10.1016/j.joi.2011.01.008

Lin, Z., Welbers, K., Vermeer, S., & Trilling, D. (2023). Beyond discrete genres: Mapping news items onto a multidimensional framework of genre cues. *Proceedings of the 17th International AAAI Conference on the Web and Social Media, Cyprus*, *17*(1), 542–553. https://doi.org/10.1609/icwsm.v17i1.22167

Loecherbach, F., & Trilling, D. (2020). 3bij3—Developing a framework for researching recommender systems and their effects. *Computational Communication Research*, *2*(1), 53–79. https://doi.org/10.5117/CCR2020.1.003.LOEC

Loecherbach, F., Welbers, K., Moeller, J., Trilling, D., & van Atteveldt, W. (2021). Is this a click towards diversity? Explaining when and why news users make diverse choices. *Proceedings of the 13th ACM Web Science Conference*, 282–290. https://doi.org/10.1145/3447535.3462506

Merten, L., Metoui, N., Makhortykh, M., Trilling, D., & Moeller, J. (2022). News won't find me? Exploring inequalities in social media news use with tracking data. *International Journal of Communication*, *16*, 1127–1147. https://ijoc.org/index.php/ijoc/article/view/17068

Möller, J., van de Velde, R. N., Merten, L., & Puschmann, C. (2020). Explaining online news engagement based on browsing behavior: Creatures of habit? *Social Science Computer Review*, *38*(5), 616–632. https://doi.org/10.1177/0894439319828012

Nguyen, M. H., van Weert, J. C., Bol, N., Loos, E. F., Tytgat, K. M., van de Ven, A. W., & Smets, E. M. (2017). Tailoring the mode of information presentation: Effects on younger and older adults' attention and recall of online information. *Human Communication Research*, *43*(1), 102–126. https://doi.org/10.1111/hcre.12097

Ohme, J., Araujo, T., de Vreese, C. H., & Piotrowski, J. T. (2020). Mobile data donations: Assessing self-report accuracy and sample biases with the iOS Screen Time function. *Mobile Media & Communication*, *9*(2), 293–313. https://doi.org/10.1177/2050157920959106

Otto, L. P., & Kruikemeier, S. (2023). The smartphone as a tool for mobile communication research: Assessing mobile campaign perceptions and effects with experience sampling. *New Media & Society*, *25*(4), 795–815. https://doi.org/10.1177/14614448231158651

Peter, J., Araujo, T., Ischen, C., Shaikh, S. J., van der Goot, M. J., & van Straten, C. L. (2024). Human–machine communication. In T. Araujo & P. Neijens (Eds.), *Communication research into the digital society: Fundamental insights from the Amsterdam School of Communication Research* (pp. 205–220). Amsterdam University Press. https://doi.org/10.5117/9789048560592_ch12

Peter, J., & Kühne, R. (2018). The new frontier in communication research: Why we should study social robots. *Media and Communication*, *6*(3), 73–76. https://doi.org/10.17645/mac.v6i3.1596

Schönbach, K. (1982). "The issues of the seventies": Elektronische inhaltsanalyse und die langfristige beobachtung von agenda-setting-wirkungen der massenmedien. *Publizistik*, *27*(1–2), 129–140.

Simon, M., Welbers, K., Kroon, A. C., & Trilling, D. (2022). Linked in the dark: A network approach to understanding information flows within the Dutch Telegramsphere. *Information, Communication & Society*. Advance online publication. https://doi.org/10.1080/1369118X.2022.2133549

Strycharz, J., & Segijn, C. M. (2022). The future of dataveillance in advertising theory and practice. *Journal of Advertising, 51*(5), 574–591. https://doi.org/10.1080/00913367.2022.2109781

Stuber, J. M., Mackenbach, J. D., de Boer, F. E., de Bruijn, G-J., Gillebaart, M., Harbers, M. C., Hoenink, J. C., Klein, M. C. A., Middel, C. N. H., van der Schouw, Y. T., Schuitmaker-Warnaar, T. J., Velema, E., Vos, A. L., Waterlander, W. E., Lakerveld, J., & Beulens, J. W. J. (2020). Reducing cardiometabolic risk in adults with a low socioeconomic position: Protocol of the Supreme Nudge parallel cluster-randomised controlled supermarket trial. *Nutrition Journal, 19*, Article 46. https://doi.org/10.1186/s12937-020-00562-8

Trilling, D., van de Velde, B., Kroon, A. C., Löcherbach, F., Araujo, T., Strycharz, J., Raats, T., de Klerk, L., & Jonkman, J. G. F. (2018). INCA: Infrastructure for content analysis. *Proceedings of the 2018 IEEE 14th International Conference on e-Science (e-Science), Netherlands*, 329–330. https://doi.org/10.1109/eScience.2018.00078

Trilling, D., & van Hoof, M. (2020). Between article and topic: News events as level of analysis and their computational identification. *Digital Journalism, 8*(10), 1317–1337. https://doi.org/10.1080/21670811.2020.1839352

van Atteveldt, W., Margolin, D., Shen, C., Trilling, D., & Weber, R. (2019). A roadmap for computational communication research. *Computational Communication Research, 1*(1), 1–11. https://doi.org/10.5117/CCR2019.1.001.VANA

van Atteveldt, W., Trilling, D., & Arcíla-Calderón, C. (2022). *Computational analysis of communication: A practical introduction to the analysis of texts, networks, and images with code examples in Python and R.* Wiley Blackwell.

van Driel, I. I., Giachanou, A., Pouwels, J. L., Boeschoten, L., Beyens, I., & Valkenburg, P. M. (2022). Promises and pitfalls of social media data donations. *Communication Methods and Measures, 16*(4), 266–282. https://doi.org/10.1080/19312458.2022.2109608

Verbeij, T., Pouwels, J. L., Beyens, I., & Valkenburg, P. M. (2021). The accuracy and validity of self-reported social media use measures among adolescents. *Computers in Human Behavior Reports, 3*, Article 100090. https://doi.org/10.1016/j.chbr.2021.100090

Vermeer, S. A. M., Kruikemeier, S., Trilling, D., & de Vreese, C. H. (2021). WhatsApp with politics?! Examining the effects of interpersonal political discussion in instant messaging apps. *The International Journal of Press/Politics, 26*(2), 410–437. https://doi.org/10.1177/1940161220925020

Vermeer, S., Trilling, D., Kruikemeier, S., & de Vreese, C. (2020). Online news user journeys: The role of social media, news websites, and topics. *Digital Journalism*, *8*(9), 1114–1141. https://doi.org/10.1080/21670811.2020.1767509

Votta, F. (2021, July 13). *Favstats/peRspective* [Original-date: 2019-05-06T21:53:53Z]. GitHub. https://github.com/favstats/peRspective

Wojcieszak, M., von Hohenberg, B. C., Casas, A., Menchen-Trevino, E., de Leeuw, S., Gonçalves, A., & Boon, M. (2022). Null effects of news exposure: A test of the (un)desirable effects of a "news vacation" and "news binging." *Humanities and Social Sciences Communications*, *9*(1), Article 413. https://doi.org/10.1057/s41599-022-01423-x

Zhao, S. (2006). Humanoid social robots as a medium of communication. *New Media & Society*, *8*(3), 401–419. https://doi.org/10.1177/1461444806061951

About the authors

Damian Trilling is a Full Professor and holds the Chair in Journalism Studies at Vrije Universiteit Amsterdam. Previously, he was an Associate Professor and member of the Political Communication & Journalism group at ASCoR. He is interested in the dissemination and consumption of news and societally relevant information in the current digital media ecosystem and in the development of methods to study them. E-mail: D.C.Trilling@vu.nl

Theo Araujo is a Full Professor of Media, Organisations and Society, and Scientific Director of ASCoR as of 2022. His research investigates the dynamic interplay between media and organisations, and what it means for society, with a special focus on trust and technology. His research has also a methodological focus on computational communication science and the implementation of large-scale data collection and analysis for communication research. E-mail: T.B.Araujo@uva.nl

Anne Kroon is an Associate Professor of Corporate Communication at ASCoR. She studies bias in digital media using computational techniques and experiments. Her research examines how social groups are portrayed in digital media and the resulting effects on prejudice and discrimination. E-mail: A.C.Kroon@uva.nl

A. Marthe Möller is an Assistant Professor of Youth & Media Entertainment at ASCoR. In her work, she analyses user-generated information in the form of comments, ratings, and reviews to study social media users' responses to and experiences of entertainment media messages. E-mail: A.M.Moller@uva.nl

Joanna Strycharz is an Assistant Professor of Persuasion & New Media Technologies at ASCoR. Combining traditional and computational methods, she studies algorithmic persuasion, its impact on consumers, their privacy, and consumer empowerment. E-mail: J.Strycharz@uva.nl

Susan Vermeer was an Assistant Professor at ASCoR and currently holds the same title at Wageningen University & Research. By combining traditional research methods with innovative and computational methods, she aims to understand the impact of the digital media environment, such as social media and algorithms, on political behaviour and attitudes. E-mail: Susan1.Vermeer@wur.nl.

16. Communication as a Social System: The Work of Loet Leydesdorff

Wouter de Nooy and Iina Hellsten[1]

Abstract

Loet Leydesdorff, professor emeritus at the University of Amsterdam, passed away on March 11, 2023, not long before the celebration of ASCoR's 25th anniversary. He joined the Department of Communication Science in 2000 and retired in 2013. As holder of the chair in Communication and Innovation in the Dynamics of Science and Technology, he represented a research programme that is rather unique within ASCoR and, more generally, within communication science. It represents a sociological approach towards communication, linking communication to meaning and innovation. In honour of Loet's memory and as a tribute to the richness of his approach and its relevance to communication science as a discipline, we briefly sketch his thinking, research programme, and impact.

Keywords: codes of communication, self-organising logic, Niklas Luhmann, triple helix, calculus of redundancy

Introduction: Codes of communication

Let us introduce Loet's thinking by contrasting it to an experiment on framing effects. In a classic framing experiment, we expose participants randomly to one of two different versions of a message, we measure some outcome variable, and compare average scores among groups. We interpret a mean difference as a framing effect.

If participants have been randomly selected from a population and the experiment was successfully conducted according to the rules, one may

1 Both authors collaborated with Loet Leydesdorff on several papers.

de Nooy, W., & Hellsten, I. (2024). Communication as a social system: The work of Loet Leydesdorff. In T. Araujo & P. Neijens (Eds.), *Communication research into the digital society: Fundamental insights from the Amsterdam School of Communication Research* (pp. 265-272). Amsterdam University Press. https://doi.org/10.5117/9789048560592_ch16

generalise the effect of the two messages to this population at this moment. Because the messages are usually not drawn at random from a population of messages, generalisation of the effect to other messages is statistically not warranted without a giant leap of faith (e.g., see Yarkoni, 2022). A typical leap of faith is to assume that a particular characteristic of the message, for example, whether it is positive or negative, has a universal meaning for human beings. This merely shifts the problem to how we define positive versus negative messages or, equivalently, how we define the populations of positive and negative messages.

The concept of populations of messages brings us to a core tenet in Loet's thinking, namely, the notion that there is an infinity of possible messages, or rather, communications, from which we only use a limited number. A question central to Loet's research programme is: How does this selection of possible messages operate and what are its effects? On the micro level of individuals, Loet accepts the notion of "double contingency" as used by, for example, the sociologist Talcott Parsons (1951): People have expectations about what other people expect from them in particular situations. Translated to the framing effects experiment, the experimenter selects two messages because these messages elicit different expectations on how receivers ought to respond to them. Participants respond in the expected way because they share the experimenter's expectation about how they should respond. From this point of view, a framing effect just "proves" that experimenter and participant communicate: They understand each other. In Loet's words, they use the same codes of communication.

According to Loet, the expectations constitute the meaning of messages. More precisely, the selections that we make from all possible communications constitute the meaning of the communications, in hindsight, because we infer expectations from preceding selections. To put it bluntly, a word derives its meaning from its context, most particularly the verbal contexts within which it is habitually used. This opens the way to measuring selections from co-occurrences.

Message selections and accompanying expectations may be experienced by individuals but they originate at an aggregated level, otherwise individuals would not share expectations, hence, attach similar meanings to messages. Building upon the work by the sociologist Niklas Luhmann (1990), Loet theorises that selections are created and operate at the societal level, more precisely, at the level of a social system. Social structure, here, is conceptualised as the structuring of communications into codes of communication. Note the fundamental role that Loet assigns to communication in society. In his view, we should investigate the structure and dynamics of message selections at the system level. Loet's notion is that "individuals do not communicate; social systems do."

This structure, however, is not static. Actually, it is constantly changing according to a self-organising logic (Luhmann, 1984; Maturana & Varela, 1980). Loet theorises an evolutionary principle for how the structuring of communications changes. In contrast to natural selection, it rests on cultural selection, in which variation, redundancy, and anticipation are central. We return to these concepts in the next section.

The most prominent example of this evolutionary process in Loet's work, which he did in collaboration with Henry Etzkowitz, is the co-evolution of universities, industry, and government as a triple helix, where research output, industry resources (notably patents), and science policy evolve over time as selections responding to, or rather, evoking selections within and across the three domains of the triple helix (e.g., Etzkowitz & Leydesdorff, 1997). Innovations are conceptualised as the realisation of not-yet-realised options within a social system. The number of not-yet-realised options can be estimated with information-theoretic indicators (see the next section) and this number can be enlarged by well-informed institutional policies.

To wind up our discussion of Loet's theoretical perspective, we would like to stress that actual persons play a minor role (if any) in his theory. Notwithstanding our self-perceptions of autonomy and self-efficacy, we behave as we are expected to behave within the social system within which we function. Communication codes have their own evolutionary dynamics, which is an important study object in and of itself.

Information-theoretic indicators

It is a challenge to empirically investigate a system, especially a self-organising, anticipatory system. Loet developed several quantitative indicators, which he applied to science as well as university–industry–government systems. At the core of his indicators is Shannon's notion of entropy (Shannon, 1948).[2] Entropy quantifies randomness or uncertainty within a system. It represents disorder and variation, absence of structure. Selections decrease variation because they foreground particular pieces of information as "signal" (meaningful information) against a backdrop of meaningless "noise." Thus, selections represent structure.

2 $H = -\sum_i p_i log_2 (p_i)$ with p_i the probability that a variable i equals 1 instead of 0.

From entropy, we derive quantitative measures of mutual information and redundancy. Mutual information is a very basic form of statistical association, namely, the extent to which we can predict one (dichotomous) variable if we know another (dichotomous) variable. In Loet's terminology, knowing one part of a system gives us expectations about other parts. Thus, we operate within domains of possible practices.

Redundancy can be seen as surplus or superfluous structure: We can predict the same outcomes from different parts (variables) of the system. Redundancy is commonly interpreted as waste, for example, repeating the same information within one message. Loet, however, has initiated a calculus of redundancy (e.g., Leydesdorff & Ivanova, 2014) in which redundancy is understood as unrealised but realisable possibilities within a system. Here, Loet builds upon the concept of mutual redundancy as redundancy among three or more variables introduced by McGill (McGill, 1954) and the concept of synergy as discussed in Krippendorff (1980) as a system's potential for realising new possibilities.

Loet demonstrated how to use mutual redundancy and synergy metrics for monitoring the development of university–industry–government relations (see above) and science (Leydesdorff, 1995). To expand on the latter, Loet argued that every scientific publication adds variation to the scientific system because it presents new data, formulates new knowledge claims, has a unique text, and so on. In contrast, citations, keywords, and abstracts in these publications are highly selective, reflecting and changing communication codes, thus opening up or closing down paths for development. Careful analysis of mutual information in citations, keywords, or abstracts reveals both how communication codes changed in the past and the likely paths they will take in the near future.

In terms of framing effects research, we could investigate how the examples of frames used as stimuli in communication research change over time, to find out how the communication science system and the media system attach meaning to particular frames. Where and when do we find high levels of synergy among research institutes or among research institutes and media outlets? Do we see a flurry of new sub-systems or specialties after times of high synergy?

Impact

Loet was very productive with 387 publications registered in the Web of Science database (Figure 16.1) and several book publications. His impact

Figure 16.1. Number of publications and citations to the publications by Loet Leydesdorff, as documented in the Web-of-Science database

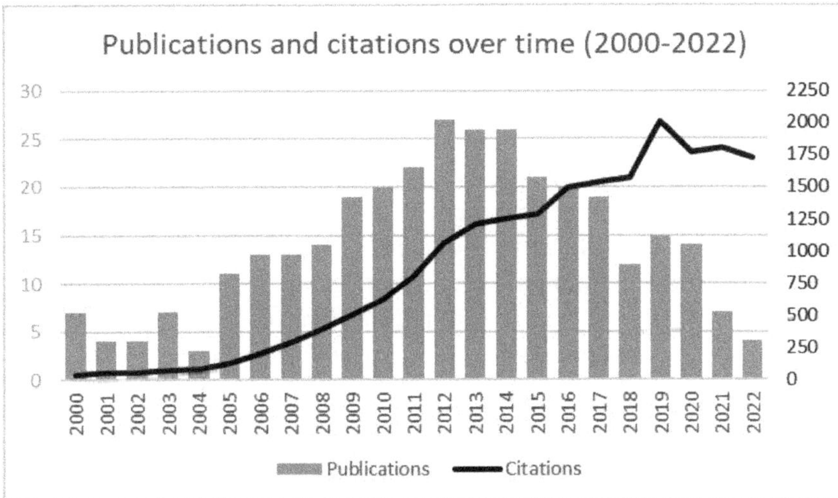

Figure 16.2. Co-occurrence Map of the Author Keywords in the Full Publication Record of Loet Leydesdorff

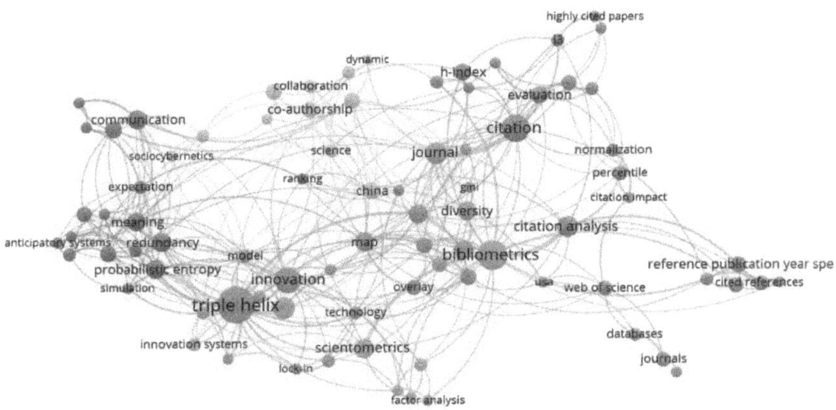

Period 1980 to 2023. Contains the 83 keywords used 3 times or more often (out of the total of 584 author keywords). Analysis and visualization in VosViewer.

within the scientific system is overwhelming with an astonishing 5,379 citations, on average, per publication. This amounted to 2,000 citations per year near the end of his life (Figure 16.1).

If we accept Loet's theory about the rise of new sub-systems as a sign of innovation, we may safely conclude that Loet has been a scientific innovator. He was one of the scientists around whom a sub-discipline emerged dealing with the triple helix of university–industry–government relations, focusing on science policy, as illustrated in Figure 16.2. In addition, the fields of scientometrics and science and technology studies flourished while he was publishing there.

Loet's practical impact within and outside academia as a social system is less easy to assess for us. It is difficult to imagine that Loet's continuous involvement in the development and critical appraisal of science performance indicators did not have any effect. For example, it seems that his pleading (with others) for the fractional counting of citations (Leydesdorff & Bornmann, 2011; Leydesdorff & Shin, 2011) has contributed to changes in the science indicators that are being used in the Netherlands. Whether his analyses of economic or scientific policies (e.g., Leydesdorff & Cucco, 2019; Leydesdorff & Fritsch, 2006) have impacted policymaking, we cannot say.

Conclusion

This is a simplified rendering of Loet's thinking and research. Loet may not have recognised parts of this summary. It is such a pity that we cannot ask him any longer! Fortunately, he published a comprehensive summary of his work that we can consult: *The Evolutionary Dynamics of Discursive Knowledge: Communication-Theoretical Perspectives on an Empirical Philosophy of Science* (Leydesdorff, 2021). What would Loet's question to communication scientists be? We think he would ask us whether we are you looking for regularities at the right level. He would try to convince us that communication is much more fundamental to human life, especially in a knowledge-based society, than the effects of a message frame on particular receivers.

References

Etzkowitz, H., & Leydesdorff, L. (1997). *Universities and the global knowledge economy: A triple helix of university–industry–government relations*. Pinter. https://papers.ssrn.com/sol3/papers.cfm?abstract_id=3404823

Krippendorff, K. (1980). Q: An interpretation of the information theoretical Q-measure. In R. Trappl, G. J. Klir, & F. Pichler (Eds.), *Progress in cybernetics and systems research, vol. VIII* (pp. 63–67). Hemisphere. https://repository.upenn.edu/handle/20.500.14332/2520

Leydesdorff, L. (1995). *The challenge of scientometrics: The development, measurement, and self-organization of scientific communications*. DSWO Press, Leiden University. https://papers.ssrn.com/abstract=3512486

Leydesdorff, L. (2021). *The evolutionary dynamics of discursive knowledge: Communication-theoretical perspectives on an empirical philosophy of science*. Springer International Publishing. https://doi.org/10.1007/978-3-030-59951-5

Leydesdorff, L., & Bornmann, L. (2011). How fractional counting of citations affects the impact factor: Normalization in terms of differences in citation potentials among fields of science. *Journal of the American Society for Information Science and Technology, 62*(2), 217–229. https://doi.org/10.1002/asi.21450

Leydesdorff, L., & Cucco, I. (2019). Regions, innovation systems, and the North–South divide in Italy. *El profesional de la información, 28*(2), Article e280214. https://doi.org/10.3145/epi.2019.mar.14

Leydesdorff, L., & Fritsch, M. (2006). Measuring the knowledge base of regional innovation systems in Germany in terms of a triple helix dynamics. *Research Policy, 35*(10), 1538–1553. https://doi.org/10.1016/j.respol.2006.09.027

Leydesdorff, L., & Ivanova, I. A. (2014). Mutual redundancies in interhuman communication systems: Steps toward a calculus of processing meaning. *Journal of the Association for Information Science and Technology, 65*(2), 386–399. https://doi.org/10.1002/asi.22973

Leydesdorff, L., & Shin, J. C. (2011). How to evaluate universities in terms of their relative citation impacts: Fractional counting of citations and the normalization of differences among disciplines. *Journal of the American Society for Information Science and Technology, 62*(6), 1146–1155. https://doi.org/10.1002/asi.21511

Luhmann, N. (1984). *Soziale systeme: Grundriß einer allgemeinen theorie*. Suhrkamp.

Luhmann, N. (1990). *Die wissenschaft der gesellschaft*. Suhrkamp.

Maturana, H. R., & Varela, F. (1980). *Autopoiesis and cognition: The realization of the living*. Reidel.

McGill, W. J. (1954). Multivariate information transmission. *Psychometrika, 19*(2), 97–116. https://doi.org/10.1007/BF02289159

Parsons, T. (1951). *The social system*. Free Press. http://archive.org/details/socialsystemoopars

Shannon, C. E. (1948). A mathematical theory of communication. *The Bell System Technical Journal, 27*(3), 379–423. https://doi.org/10.1002/j.1538-7305.1948.tb01338.x

Yarkoni, T. (2022). The generalizability crisis. *Behavioral and Brain Sciences, 45*, Article e1. https://doi.org/10.1017/S0140525X20001685

About the authors

Wouter de Nooy was appointed at the Department of Communication Science at the University of Amsterdam from 2007 until 2023, first as an Associate Professor in Research Methods, later as a Senior Lecturer. E-mail: Wouter@wdenooy.nl

Iina Hellsten worked as a Marie Curie Postdoctoral Fellow (2002–2004), and as an Associate Professor (since 2016) at ASCoR. She has collaborated with Loet Leydesdorff since 2002 on several co-authored articles, first on semantic networks, and more recently on the innovative "n-mode" network approach. E-mail: I.R.Hellsten@uva.nl